D0996384

ROBOTS

IN FACT **AND FICTION**

ROBOTS

IN FACT AND FICTION

BY MELVIN BERGER

FRANKLIN WATTS
NEW YORK | LONDON | TORONTO | SYDNEY | 1980

Photographs courtesy of Universal Studios: pp. 44, 46, 49, 50, 53, 62, 68 (top), 74; Union: p. 40; Gold Key Comics: p. 64; UFA: pp. 58, 60; Marvel Comics: p. 63; New York Public Library Picture Collection: pp. 11, 42; Metro-Goldwyn-Mayer, Inc.: pp. 68 (bottom), 76, 77, opposite title page; American International: pp. 3, 55, 56; Paramount Pictures: pp. 72, 73.

Photograph opp. title page is
from the film *Logan's Run.*

Library of Congress Cataloging in Publication Data

Berger, Melvin.
Robots in fact and fiction.

Bibliography: p.
Includes index.
SUMMARY: Discusses a variety of robots, real and imaginary, with or without human form, which perform some humanlike functions—seeing, hearing, thinking, speaking, etc.
1. Automata—Juvenile literature. 2. Robots, Industrial—Juvenile literature. [1. Automata. 2. Robots] I. Title.
TJ211.B47 629.8'92 80-14139
ISBN 0-531-04155-7

Printed in the United States of America
5 4 3 2 1

CONTENTS

Many people and organizations helped me in the writing of this book. I am grateful to them all. In addition, though, I would like to offer particular thanks to Daniel S. Giroux (Argonne National Laboratory), Cynthia M. Outcalt (Bell Laboratories), Roy Cooper (Burden Neurological Institute), John G. Anderson (Citibank), Egyptian Government Tourist Office, Nancy E. Jamison and Peter Van Avery (General Electric), Uta Hoffman (German Information Center), Walter B. Hendrickson, Jr., Robert L. Meyer (Hughes Aircraft Company), F. Harchini (Italian Government Travel Office), Frank Bristow and William M. Whitney (Jet Propulsion Laboratory), Sharon L. Bronkema and G. S. Martense (Lear Siegler Automated Systems Division), Dr. Michael Gordon and Helen Webb (University of Miami), William Der Bing (NASA), Jerry Ohlinger, Anthony Reichelt (Quasar Industries), Al Pinsky and Frank J. Strohl (RCA), John F. Young (Robotics Laboratory, University of Aston), Benjamin Skora (Skora Enterprises), Erika Faisst (Swiss National Tourist Office), Sherry Angle (Twentieth Century Fox), J. F. Engelberger and Ellen Mohr (Unimation), Jack S. Schneider (U.S. Department of Energy), Roy L. Morrow (Westinghouse), and Donald Gallup (Yale University Library).

ROBOTS

IN FACT AND FICTION

CHAPTER 1

ROBOTS: FACT OR FICTION?

A robot is a "smart" machine. It is a mechanical device that can carry out some very humanlike functions—seeing, hearing, thinking, speaking, and so on. A robot is much like a human being, even when it is not shaped like one. Someone without willpower, or a person who cannot think for himself or herself or feel emotions, can be said to act like a robot.

Some robots exist in fact. These real, working robots are found in homes, businesses, factories, laboratories, on missions in outer space, and in many other places.

Other robots exist only in fiction. These make-believe robots are the wonderful, fanciful objects found in films, on television, and in books and plays.

Can you tell which of the following are robots of fact and which are robots of fiction?

- A giant robot steps out of a spaceship that lands on the White House lawn. It melts all the guns and weapons the guards and soldiers turn on it.

- A female robot organist bows to the audience, takes a deep breath, flexes her fingers, and begins to play. During the performance, she sways to the music and sighs deeply several times. She plays five different pieces.
- A chess-playing robot beats such worthy opponents as Benjamin Franklin, Napoleon Bonaparte, and Frederick the Great.
- A domestic robot cooks and serves the meals, cleans the house, takes out the trash, and even walks the dog.
- A 25-foot (7.5-m) tall robot walks over rough, hilly terrain on its 16-foot (4.8-m) long legs.
- A robot scientist lands on Mars. It measures the temperature and humidity, and tests the soil for signs of life.
- A surgeon creates a robot out of human corpses.
- Robots working on an assembly line put together a car from separate pieces of metal and glass.
- Robots in a resort amuse the guests by helping them to act out their dreams.

It is sometimes hard to tell robots of fact from robots of fiction. They often resemble one another. Robots of fact inspire writers to create robots of fiction. And robots of fiction move inventors and scientists to develop more advanced robots of fact.

As you read through this book, you will discover the truth about each of the robots just described. You may be surprised to learn which are robots of fact and which are robots of fiction.

ROBOT LINGO

Robotics is the science of robots. Scientists in this field are working to build improved robots and to find new ways for robots to help humanity.

Many robot designers feel that since robots are supposed to *act* somewhat like people, they should also *look* somewhat like people. Also, since real robots often do the work of humans, people-sized robots are generally thought to fit the environment better than robots of other shapes and sizes.

A robot in human shape or form is usually called an *android,* or *droid* for short. Sometimes it is called a *humanoid.*

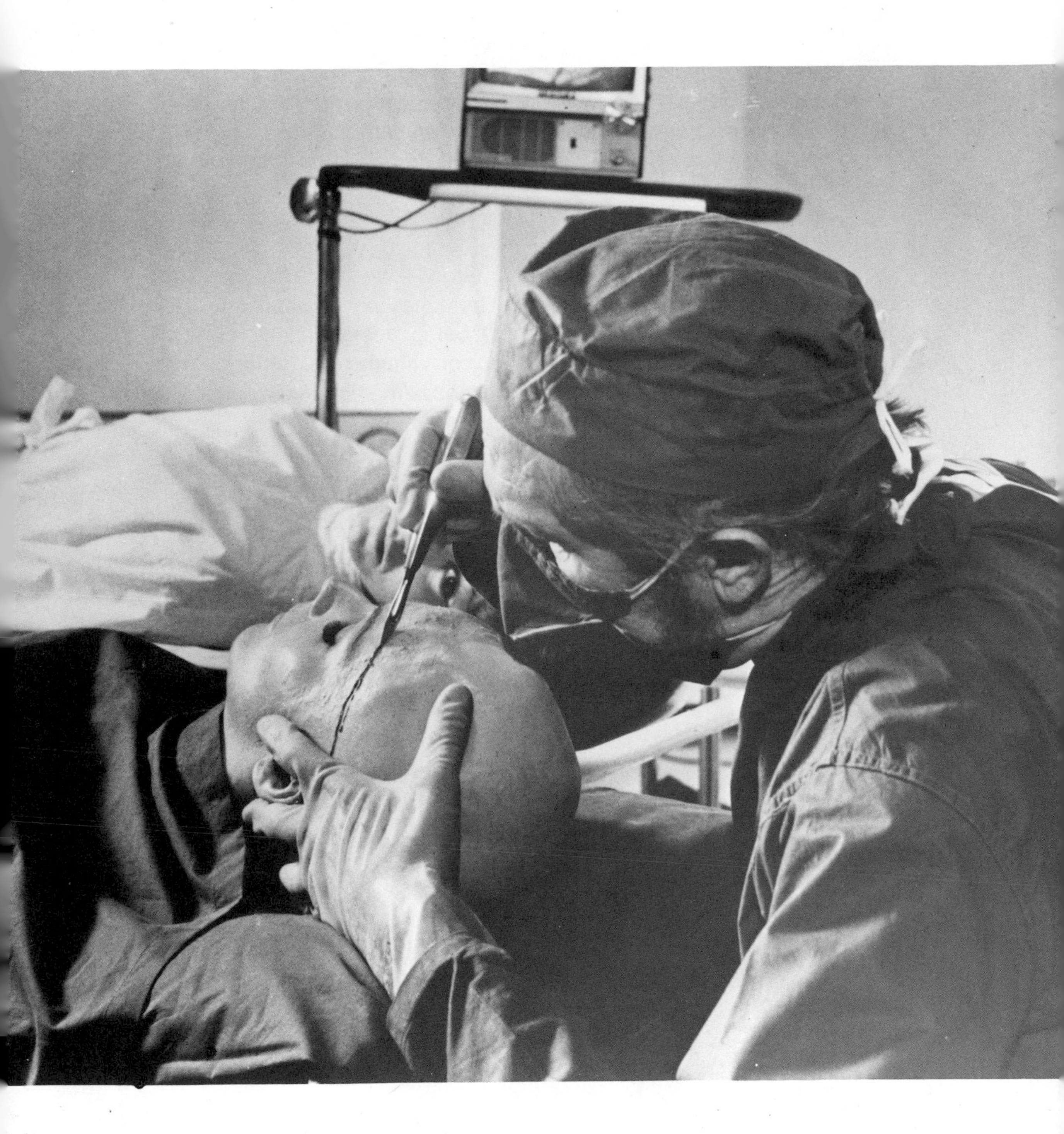

In the classic film *Scream and Scream Again*, a mad doctor creates living robots.

The word *cyborg* is a combination of *cyb*ernetics (the science of automatic control and communication), and *org*anism (a living being). A cyborg is part robot and part human being.

A *drone* is a robot that cannot act on its own. It is totally under the control of a human operator.

An *automaton* is a mechanical figure. The main feature of an automaton is that it is self-powered.

Androids, cyborgs, drones, automatons—we live at a most exciting time in the history of robots. There has never been more interest in the subject. Real robots are being designed and built all over the world. And the imagination of filmmakers, writers, and television producers are turning out truly dazzling fictional robots.

In fact and in fiction, the age of robots is here!

CHAPTER 2

MECHANICAL ROBOTS

From about 1500 until 1900, people were fascinated by mechanical robots, or automatons. These figures could make only a limited number of movements. They were usually operated by clockwork mechanisms. In fact, many of the early automatons were connected to large outdoor clocks in European cities.

The automatons on the clock tower in the Piazza San Marco in Venice, Italy, were built in 1497. Since then, the two giant robots have been raising their hammers, turning, and striking the large bell every hour of the day. They are operated by a mechanism hidden within the tower.

The mechanical robots around the clock on the Frauenkirche (Church of Our Lady) in Nuremberg, Germany, are even more elaborate. Every time the clock strikes, dozens of figures march around, while others play bells, horns, drums, and other instruments.

By the 1600s, some automatons were being made that had nothing to do with clocks. They were part of the growing effort to picture nature in mechanical terms. Scientists and philosophers constructed machines that imitated different human and animal actions.

Above: the clock-tower robots in Venice's Piazza San Marco strike the hour as usual. Right: just beneath the clock of the Frauenkirche in Nuremberg, Germany, are dozens of robots that march around and play musical instruments.

One of the first mechanical robots was said to have been built by René Descartes around 1650. Descartes was a brilliant French mathematician and philosopher and a leading figure of the Age of Reason. In his writings he stressed the separation of body and soul. The body without a soul, he said, is like a highly complex machine. This set the philosophical basis for creating lifelike automatons.

Most automatons were powered by springs. Winding them tightened the springs. Then, as the springs uncoiled, they made the robots move.

Descartes built a robot that was a perfect likeness of a beautiful young woman. He called her Francine. Not only did Francine look real, but she had a mechanism that let her move the way people do.

Descartes went on a sea voyage and took Francine along in a large packing case. On the crossing, a fellow passenger became curious about the case. He sneaked a look inside. When he saw Francine, he reported it to the captain of the ship. One look at the robot convinced the captain that Francine was the work of the devil. He ordered her thrown overboard. Francine was never seen again. And Descartes never built another automaton.

Actually, there is no real proof that Descartes built Francine. But the story is generally accepted as part of the factual history of robots.

During the 1700s, wealthy Europeans often ordered elaborate, lifelike robots to be custom-made for themselves. Perhaps the most amazing robots built during this period were those by the father and son team of Pierre and Henri-Louis Jacquet-Droz. One automaton is a boy seated at a desk holding a pen in his hand. The boy writes a message, and when it is finished his hand goes back to dot the i's and cross the t's.

This automaton, or mechanical doll, is called the Draftsman. It was built in 1744, and it still writes messages today for visitors to a museum in Switzerland.

Another remarkable Jacquet-Droz automaton is a young female organ player. When performing, the organist first glances at the audience and then bows shyly. Next, she lowers her eyelids, turns to the instrument, wriggles her fingers to warm up, and begins to play. Swaying to the music, she works her way through the five pieces she knows. Throughout the performance her chest rises and falls as she breathes deeply. At the end, she stands up, nods, and curtsies.

The only eighteenth-century robot more famous than the Organ Player was the Turbaned Turk. The Turk was a chess-playing automaton that Baron Wolfgang von Kempelen constructed in 1783. Although it proved to be a hoax, crowds kept coming to view this amazing chess whiz until it was destroyed by fire in 1854.

The Turbaned Turk was dark, swarthy, and slightly larger than life. He wore a plumed headdress and fur-trimmed robes. In his left hand he carried a long-stemmed Turkish pipe.

The figure sat behind a large cabinet with doors and a drawer. A chessboard rested on top of the cabinet. Before each performance, Baron von Kempelen parted the Turk's robes to show the audience the mechanism that operated the robot. He opened the doors and pulled out the drawer so that people could see inside.

The human player removed the pipe from the Turk's left hand to start the game. The Turk made the first move, and the game was on. The Turk always won. He beat Benjamin Franklin, Napoleon, Frederick the Great, George III, and Louis XVIII among many, many others. The only game that he is known to have lost was to Empress Maria Theresa of Austria. (And it is believed that he lost this game on purpose.)

After touring for many years in Europe, the Turbaned Turk was brought to America. In May 1827, the Turk was shown in Baltimore.

The most famous of the mechanical robots was the Turbaned Turk. In time, though, he was found to be a fake, operated by a man hidden in the cabinet.

1
2
3

Two boys hid on a roof to catch a glimpse of the robot. They saw much more than they expected. Unaware of their presence, the Turk's owner opened the cabinet. Out stepped a short, stooped man!

The news quickly spread that the Turbaned Turk worked, not by a brilliant mechanism but by an expert chess player crammed inside the cabinet. The chess expert had avoided detection earlier by always hiding from view as each door and drawer were opened in turn. According to one story, the Baron first built the Turk to help a legless Russian chess master to escape his native country after he was accused of political crimes.

The widespread disappointment in learning that the Turbaned Turk was a fake helped bring the interest in mechanical robots to an end. Also, there was a problem in powering the robots. The robot makers had gone as far as they could go with spring-driven mechanisms. They needed to find a better source of energy.

CHAPTER 3

ELECTRICAL ROBOTS

By the end of the 1800s electricity was becoming widely available. Robot makers took advantage of discoveries in the field of electricity to build a generation of electrical robots, today's robots of fact.

DYNAMO

To help us understand how these robots work, let us imagine a typical electrical robot. We'll make it an advanced general-purpose robot, one that might be used as a servant around the house. Its name will be Dynamo.

Dynamo is an android. It looks like a human being with a shiny metal skin. The electricity that supplies the power for Dynamo comes from a battery hidden inside the body. Miles of wire, also inside the body, connect dozens of electrical devices. Altogether they make up the robot's electrical circuit.

The way in which the circuit is put together, its programming, determines how Dynamo acts and what it can do.

Dynamo receives information from the outside world by means of sensors. Sensors are devices that detect a variety of physical properties, such as light, sound, pressure, or heat. They can be used to perceive their environment and react appropriately to it.

Dynamo's eyes are sensors for detecting light. They are made of photoelectric cells. When light strikes a photoelectric cell, it creates a small electrical current. The amount of light determines the flow of current.

Dynamo can be programmed to use this electricity in various ways. It may be programmed to move toward a light or away from it. Or, it may be programmed to turn on a lamp if there is not enough light in a room.

Dynamo has microphones for ears. Microphones change sound vibrations into patterns of electrical pulses. The patterns of electricity are then used to follow orders such as "Go forward!" "Stop!" "Turn around!" that may be given by voice or sound signals.

The robot has an electric thermometer that measures temperature. The higher the temperature, the greater the flow of electricity.

Dynamo senses pressure by means of a switch. When there is no pressure the switch is open and there is no flow of electricity. Pressure pushes the switch closed, and electricity flows.

To measure pressure, Dynamo has a strain gage. A strain gage produces an electrical current when it is pressed. The greater the pressure, the greater the flow of electricity.

Robot speech is usually produced by means of short lengths of prerecorded tape and a loudspeaker similar to that found in a radio, television, or record player. Each length of tape contains a separate word, phrase, or number. The electrical circuit is programmed to play back the tapes in a certain order. This produces intelligible speech.

Basically, Dynamo's brain is a microcomputer. It can calculate and solve problems and remember vast amounts of information. It can take information from the outside, compare it with information already in its memory, and decide what to do next.

Because it uses microprocessors, the computer is small enough to fit inside the robot's body. Microprocessors are tiny chips of silicon programmed to do calculations that at one time required a room-sized computer.

The motors that move the various parts of Dynamo are powered by electricity. Gears, wires, and metal rods push or pull the movable parts of the robot. Liquid under pressure (hydraulics) or air under pressure (pneumatics) can also be used to move the robot's limbs, head, or body.

Finally, Dynamo has a feedback system. This system is connected to the various movable parts of the robot. Feedback lets the robot know where it is and, if it is not in the correct position, how to get where it should be.

Let us say that Dynamo is programmed to pick up a pencil. The robot's arm reaches out toward the item but misses. The feedback system senses whether the robot has reached too far or not far enough. This information is fed back to the robot's motors. They then move the hand until it is on the pencil.

What makes Dynamo a true robot and not just an advanced machine?

Dynamo has several features that are seldom built into machines. Sensors bring it information from the environment. A computer allows it to make decisions on the basis of this information. The feedback system helps it achieve its programmed goals. And it does all of this without human control or direction.

EARLY ELECTRICAL ROBOTS

The first electrical robot made its appearance at the 1929 International Radio Exhibition in Paris, France. It was an electrical robot dog invented by Harry Piraux. Piraux named his creation Philidog.

Philidog, with its clever arrangement of photoelectric cells and motor-driven wheels, could do two outstanding tricks. It could follow the path of a flashlight that was moved in front of it. And it could bark when the light came too close.

Over the next few years several electrical androids were built, mostly for fairs and exhibitions. They were programmed to do a number of tricks; they could move and speak, answer some questions and solve some problems, follow spoken commands, light a candle, shoot a gun, tell the time, and shuffle cards.

Two robots that Westinghouse Corporation built for its exhibit at the 1939 New York World's Fair stand out in this generation of robots. One was Elektro, a male android. The other was Elektro's companion, a robot dog named Sparko.

Elektro was 7 feet (2.1 m) tall and weighed 260 pounds (118 kg).

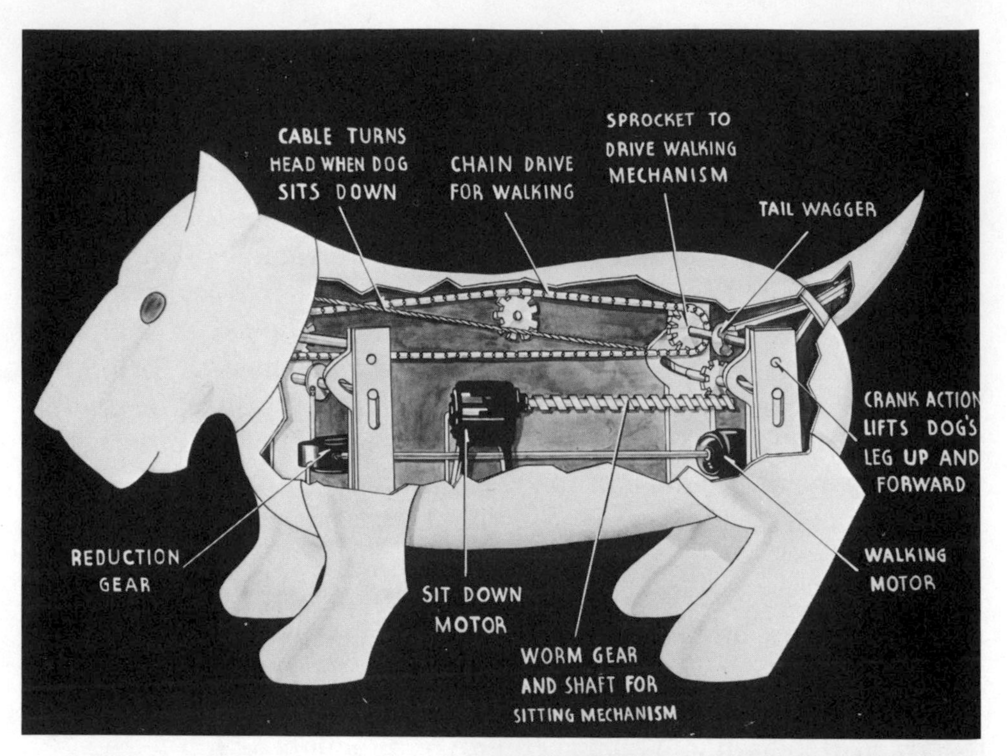

Sparko, a robot dog, was a companion to Elektro. It could walk, bark, sit, and follow a source of heat.

Elektro was an electrical robot built for the New York World's Fair in 1939. He could walk, move his arms, smoke a cigarette, and count, among other things.

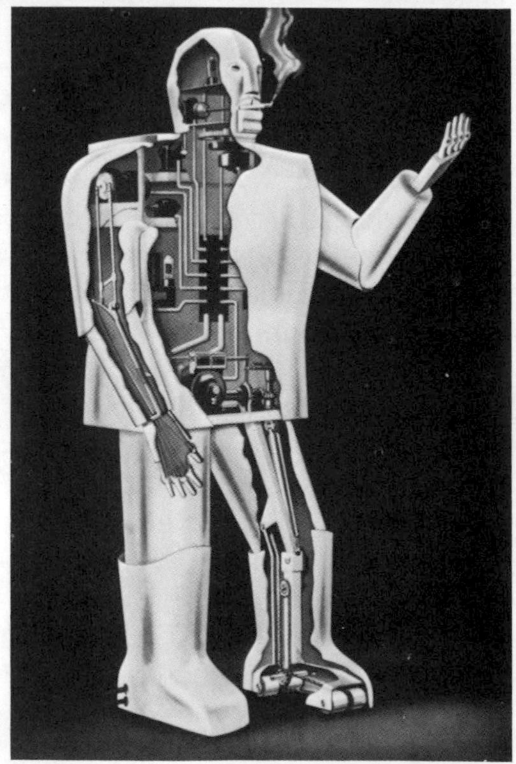

He contained eleven electrical motors that let him do some twenty-six different tricks. These included walking forward and back, moving his arms and fingers, smoking cigarettes, counting on his fingers, saluting, and recognizing as well as naming the shades red and green.

Sparko was similar to Philidog in that it was attracted to light. In addition it had a heat sensor that also attracted it to heat. It made a big hit at the fair because it followed passersby, attracted by the heat of their legs.

Unfortunately, Sparko met a tragic end. One evening the door to the Westinghouse exhibit was left ajar. As a car drove by, Sparko, attracted by both the light and heat of the passing vehicle, dashed out of the building and ran toward it. The dumbfounded driver did everything he could to avoid the determined dog. But the sad result was that the robot was destroyed.

ROBOTS IN RESEARCH

During the 1940s World War II took all of the time and energy of scientists. Between working for the war effort and helping to rebuild after the war, they had little opportunity for speculative robot research.

In the 1950s, though, robotics entered a new period of activity. Dr. W. Grey Walter, a scientist at the Burden Neurological Institute in Britain, was one of the pioneers. He used robots to help him in his research on the human brain.

Dr. Walter built small, simple robots that could move, see, hear, feel pressure, and think. To enable them to move, the robots were equipped with three wheels arranged like a tricycle. They had two photoelectric cells for eyes and a microphone for ears. They sensed pressure through a movable outer shell that was attached to a number of strain gages. An electrical circuit served as the robot's brain. The robots were powered by a rechargeable battery. Because of the way they looked, everyone began to call the robots "turtles."

Dr. Walter set out to "teach" his turtles to go toward a light. He programmed them to look for a light if one was not seen. He also instructed them to go around any obstacles in their way.

Visitors to Dr. Walter's laboratory could often observe the distinguished scientist on his hands and knees with one of the robot turtles. First, the turtle would glide silently across the lab, heading straight for a lighted lamp that Dr. Walter had placed on the floor.

Then Dr. Walter would shut the lamp and turn on another one far off to the side. In response, the turtle would pause, turn, and head for that lamp.

If Dr. Walter placed a book on the floor in front of the turtle, the turtle would slam into it—at first. But after a moment's pause it would turn and go forward again at a slightly different angle. Again a collision, and again the turtle would stop, turn, and try to get past. After several tries it would bypass the obstacle and head toward the light.

Dr. Walter also programmed his turtles to stop for a moment if he blew a whistle and then set out again toward the light. Later he made the instructions even more complicated. The robots were to go forward when they heard a whistle, just as they did when they saw a light.

In Dr. Walter's turtles, the electric eye ran the drive motor. When the photoelectric cell faced a light, the flow of electricity made the drive motor move the turtle forward toward the light. Absence of light started the scan motor, which turned the turtle about in search of light.

The turtle's microphone ear was also connected to the drive and scan motors. The sound of the whistle by itself turned off both motors. That made the turtle stop for a moment.

The turtle's movable shell contained strain gages. When the turtle bumped into something, the flow of electricity temporarily shut off the drive motor. The scan motor, though, kept on working. Thus, when the drive motor started again a few seconds later, the turtle would move forward in a slightly different direction.

Around this same time, Dr. Claude Shannon of Bell Telephone Laboratories built Theseus, a robot mouse, for his research. Dr. Shannon taught Theseus to find its way through a maze of twenty-five squares or boxes in the shortest possible time.

In the beginning, Theseus tried one direction after another in each square it entered, until it found the opening. Then it went to the next square. By this trial-and-error method, it took Theseus two minutes to get to the final square. But Theseus was programmed to learn from its mistakes. It took Theseus only fifteen seconds to get through the maze the second time.

Robot turtles and mice contributed to our understanding of human learning. They also made it possible for future robot makers to add learning ability to their robots.

CHAPTER 4

ROBOTS AT WORK

DRONES

Drones are robots that work under the control of human operators. They are the simplest robots made today. Sometimes called master/slave manipulators, drones are really mechanical extensions of the human body.

The first drones were created in a nuclear research laboratory in Illinois. Workers in this lab had the dangerous job of handling radioactive materials. Exposure to these materials can cause serious illness or even death. The only protection is either a concrete wall or a sheet of lead.

The radioactive materials were kept in special rooms with thick concrete walls. These rooms, called hot cells, were equipped with master/slave manipulators that reached into the room through sealed openings in the wall. Human workers sat behind the concrete walls, peered into the room through windows of leaded glass, and operated the manipulators. As the workers moved their fingers and arms, the metal fingers and arms of the manipulator moved in exactly the same way inside the hot cell. Thus, only the mechanical extensions ever came in contact with the dangerous materials.

Master/slave manipulators were created to handle dangerous radioactive materials. Here a worker uses a manipulator to examine a nuclear reactor.

Master/slave robots can handle the most fragile laboratory glassware. They can also operate the heavy machines used to repair nuclear reactor parts. They can work in temperatures above boiling or below freezing. They can tolerate the most poisonous gases.

Drone robots are also being used to do work on the ocean floor. Basically, the underwater drones are powerful mechanical arms. They can be controlled either from a ship on the surface or by an operator inside the robot itself.

Beetle is the largest drone robot in the world. It is about 25 feet (7.6 m) tall and weighs 85 tons (77 m.t.). Its arms are 16 feet (4.9 m) long. The operator rides in a cab with 1-foot (30-cm) thick lead walls and a 2-foot (60-cm) thick window of leaded glass. Beetle moves around on giant tank treads.

Beetle has television camera eyes that are held at the end of its arms. These eyes enable Beetle to see around corners and peer into dangerous places. Beetle is used in the building of nuclear-powered rockets and is kept available for rescue missions around nuclear materials.

CAM (*C*ybernetic *A*nthropomorphic *M*achine), a smaller, more advanced robot than Beetle, was built for the U.S. army. Instead of treads, CAM uses its powerful legs to move across terrain that might stop Beetle.

CAM has two arms and two legs, each one about 6 feet 6 inches (2 m) in length. The operator, seated in the cab, controls the robot's arms and legs by moving his or her own arms and legs. Each time the operator moves a limb, the robot moves a limb. But the robot's movements are four times as big and four times as powerful. CAM can walk as fast as 35 miles (56 km) an hour.

CAM can climb steps and walk over obstacles as high as 4 feet (1.2 m). It can make its way through narrow openings and avoid large objects in its way. Its arms are strong enough to pick up a 500-pound (227-kg) weight and carry it for long distances without tiring. It once lifted the front end of a jeep and pushed the car along with one foot.

A walking robot similar to CAM was developed a few years ago for use in exploring the moon. This vehicle has eight legs. It can walk over rocks or through soft sand.

A smaller version, which uses a motor from an electric drill and batteries from a motorcycle, is now used in a "walking" wheelchair for physically disabled persons. The vehicle is controlled by a

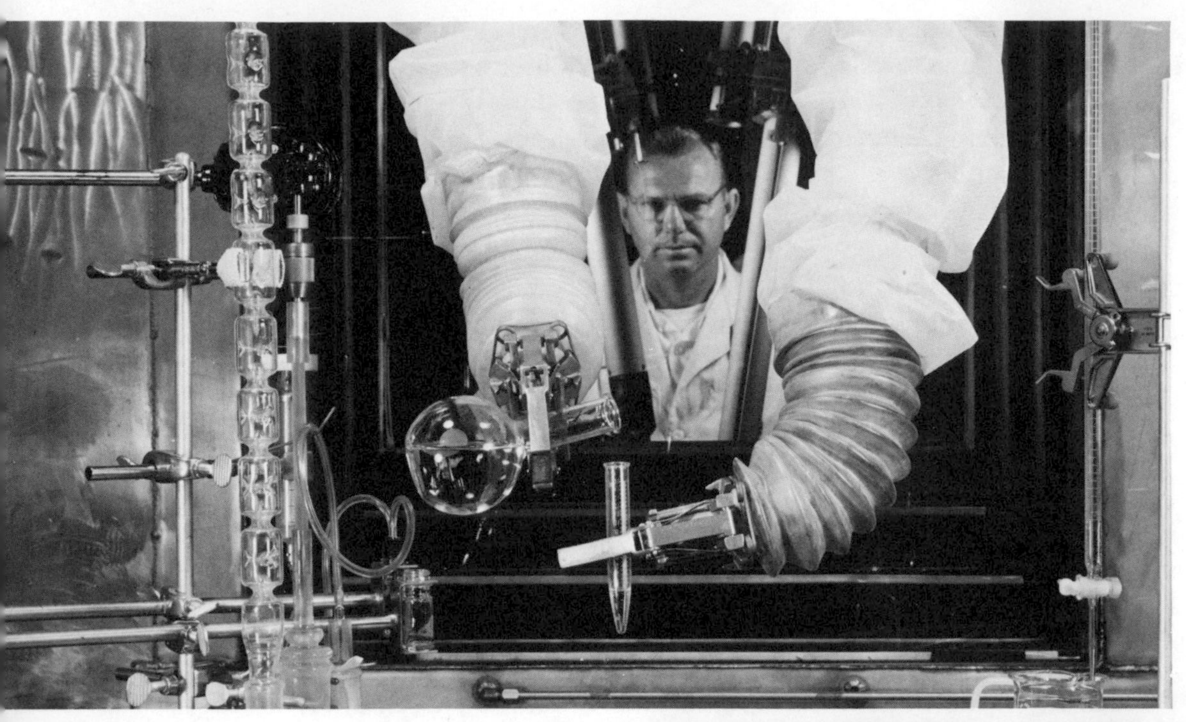

In the hands of a skillful operator, a master/slave manipulator can handle delicate laboratory glassware without breaking it.

The operator actually rides in this underwater drone robot. He or she can look out through the large round window.

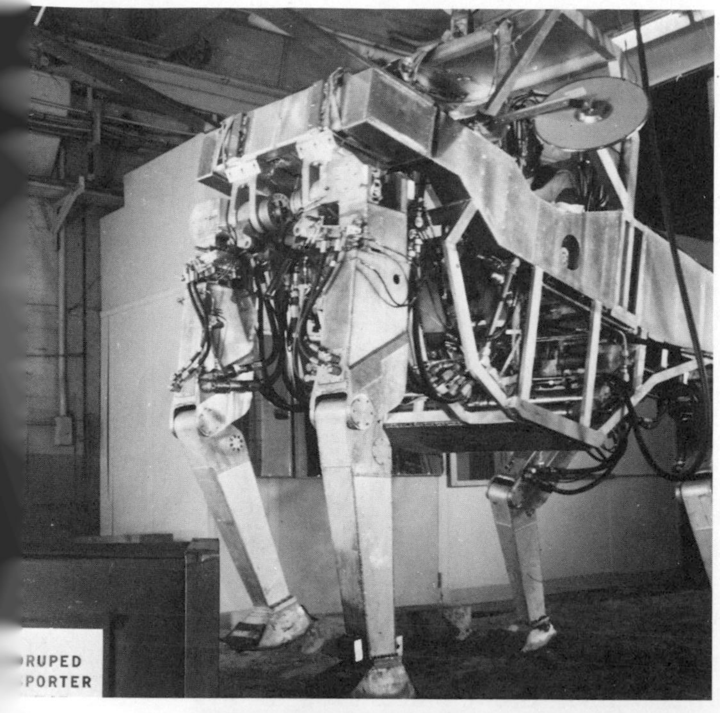

CAM is a four-legged drone robot, controlled by an operator seated in the cab. CAM can climb over obstacles and carry heavy loads.

This walking wheel-chair for the physically disabled is a drone robot based on plans developed for a vehicle to explore the moon.

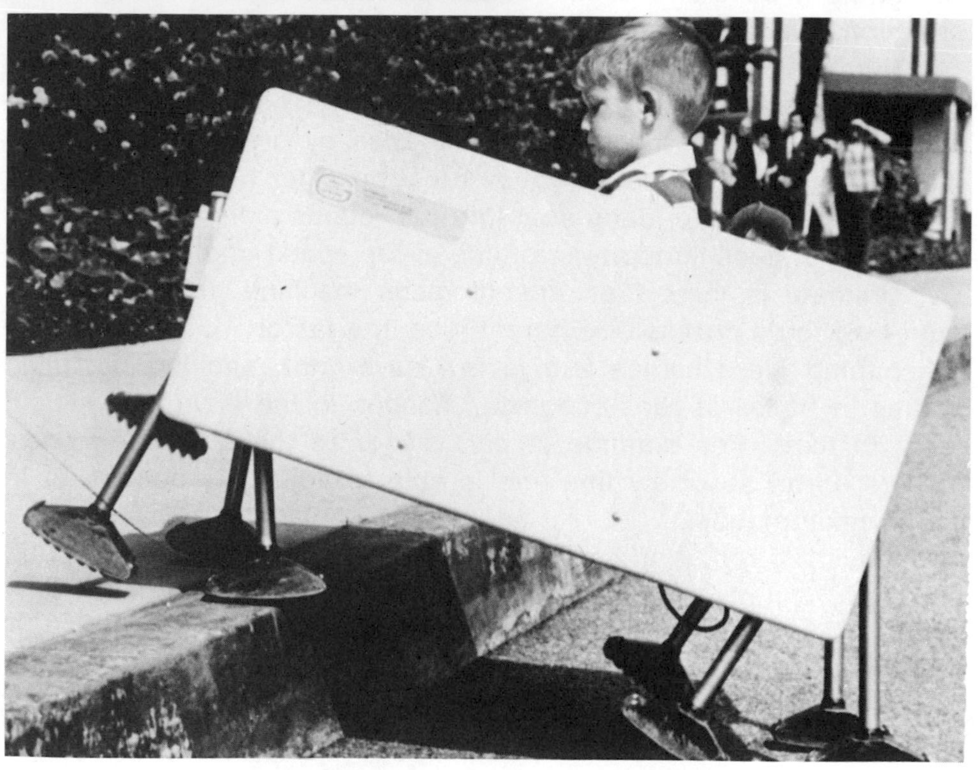

single stick that the user pushes in the direction he or she wants to go. For people who cannot use either their hands or legs, the stick can be equipped with a chin cup, making it possible to steer the vehicle with the head.

INDUSTRIAL ROBOTS

In the past, workers were forced to do many jobs that were dangerous, difficult, or unpleasant. Some factory workers still do. In one plant, employees dip metal parts into baths of hot acid. In another, workers feed heavy steel plates into giant presses. Some bakers handle breads around ovens that reach temperatures of several hundred degrees.

Now, more and more of these kinds of tasks are being performed by industrial robots. These robots can often work faster, cheaper, more safely, and more efficiently than human factory workers.

Unimate is a popular industrial robot. It is made of two parts, a powerful mechanical arm and hand that can move and bend in six different ways and a built-in computer brain that stores all the directions Unimate needs to do many different jobs.

Putting Unimate to work in the factory is quickly done. The users can train the robot in a matter of hours. They move it through the various steps of the required operation by pressing buttons on the teach-control device. Then, by pressing other buttons, they lock in the instructions for each step into the robot's memory. From then on the robot goes through its routine at top speed and without error.

Unimate is more than an automatic machine. It can be programmed to do dozens of different jobs in a factory. One day it may be shaping glass bottles and, a few days later, packing the filled bottles in boxes. It can recognize changes in the environment and react to them. For example, it can recognize the different models of cars on an assembly line and is able to adjust its actions to fit the particular model.

Industrial robots are at work in thousands of factories around the world. In plants that make cars, robots stamp out individual car parts, weld them together, and perform a number of other jobs along the assembly line. They may work side by side with other industrial robots, each one doing its own job. In factories that produce glass, metal, and plastic products, robots do everything from

When materials must be handled at very high temperatures, an industrial robot such as Unimate (top) can be used. Below: a line of industrial robots at work in an automobile factory. They can recognize the different models of cars and adjust their actions accordingly.

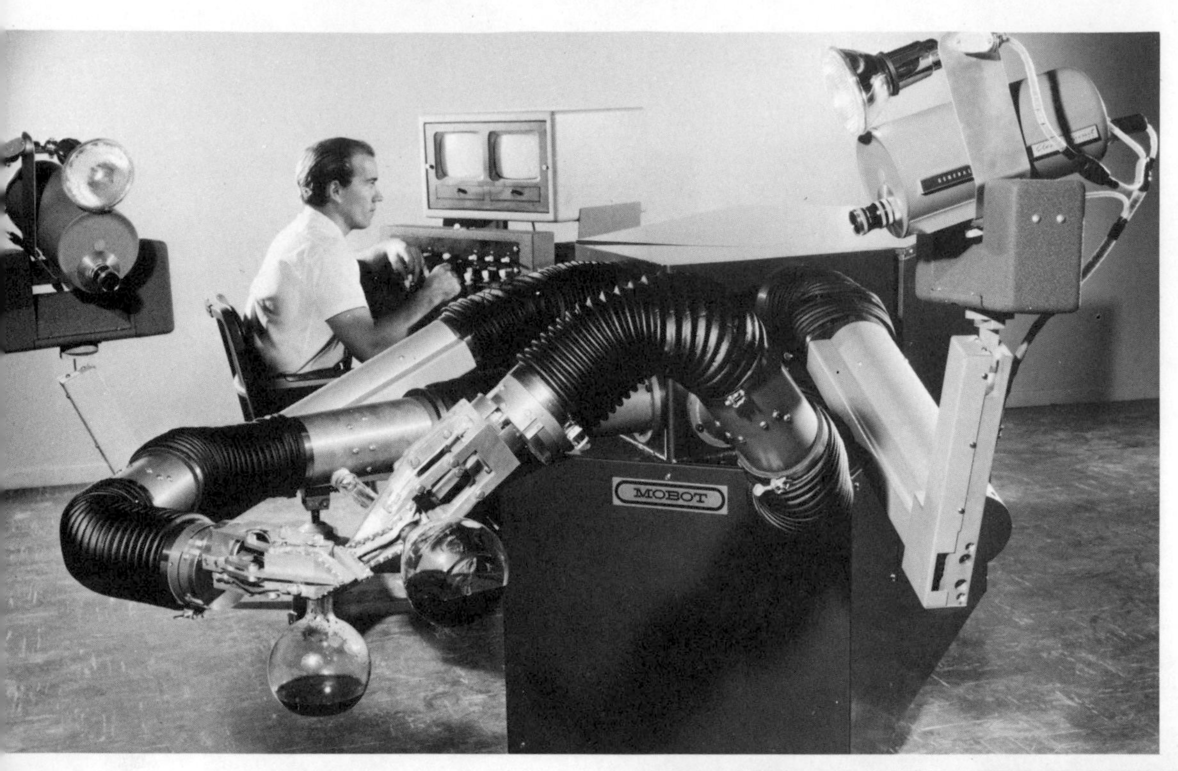

Mobot Mark II is
a mobile, industrial
drone robot controlled
by an operator
seated at a television
screen that shows
what Mobot is doing.

The Mailmobile silently
glides through
large offices carrying
mail, papers, and
small packages from
person to person.

loading items into red-hot furnaces for heat treatments to packing the finished items into boxes for shipment.

Most industrial robots are bolted down in place. The Mobot Mark II, though, is a mobile industrial robot. It moves to the work instead of waiting for the work to be brought to where it is located. The Mobot is controlled by an operator who sits at a control panel with two television screens. Two television cameras on Mobot help the operator to follow Mobot as it moves about.

One moving robot already widely used is the Mailmobile. This wonder is a 4-foot (1.2-m) tall vehicle that delivers mail, papers, and small packages in office buildings. The battery-operated robot moves along an invisible chemical pathway sprayed on the floor or carpet. An ultraviolet light beneath the robot makes the pathway glow. A light sensor guides the robot along the glowing path at 1 mile (1.6 km) an hour.

The Mailmobile can make any number of stops along its path. Each time it stops, it pauses for twenty seconds to allow people to remove or add mail. At the end of the twenty seconds, the Mailmobile gives a warning beep and then proceeds to the next stop.

The Mailmobile's bumpers are sensitive to the slightest pressure. If the robot bumps into someone or something in its path, it stops at once. Some Mailmobiles are equipped with light sensors. They stop automatically when approaching an obstacle.

Industrial robots can be adapted for tasks other than work. One robot appeared on a television talk show. While on the air, it conducted the orchestra, hit a golf ball, and opened a can of soda and poured it into a glass.

A factory owner, afraid that his workers might resent the new robot he had bought, arranged to have the robot serve them coffee one morning. When the robot continued pouring, even though there were no more cups, the workers realized that there was little immediate danger they would be replaced by robots.

SPACE ROBOTS

Viking I and Viking II are a pair of space robots that landed on the planet Mars in 1976. The two robots conducted an extensive series of scientific tests. They measured the Martian temperature, humidity, air pressure, and wind speed. They were on the alert for any tremors or earthquakes. They tested the strength of the soil, estimated how tightly the clumps of soil were bound together, and used a magnet to search for iron in the soil. In addition, they scooped up samples of soil and ran it through five different biological tests, looking for any signs of life. (Although the first results indicated that there might be life, the final conclusion was that there is no life on the planet.)

The second generation of space robots now being built will be more advanced than the Viking robots. They will have several hundred thousand instructions in their memories. They will be mobile and able to move about on the surface of a planet without falling into any holes or striking any boulders. Also, they will be able to locate and pick up objects on the scientists' commands.

The new space robots will have two television cameras and a laser beam, all of which will be able to turn in any direction. An arm will reach out to pick up objects. Sensors in the "fingers" at the end of the arm will shine tiny beams of light. The time it takes for the light to be reflected back will be a measure of how close the fingers are to their target. The fingers will have switches to shut them in a tight clasp once the object is reached.

MEDICAL ROBOTS

Several times a week a patient named George has a heart attack at the Medical School of the University of Miami. Each time he becomes ill, the medical students examine George. They check his heartbeat, pulse, temperature, and breathing, and listen for sounds from his chest. They decide what disease he has and how it should be treated.

This full-sized model (opp. top) of the Viking space robot shows the exact position of the real robot on the planet Mars as it conducted various scientific experiments. Models of new space robots (right) are now being built and tested. They will be able to move about and do many more tasks than the Viking robots.

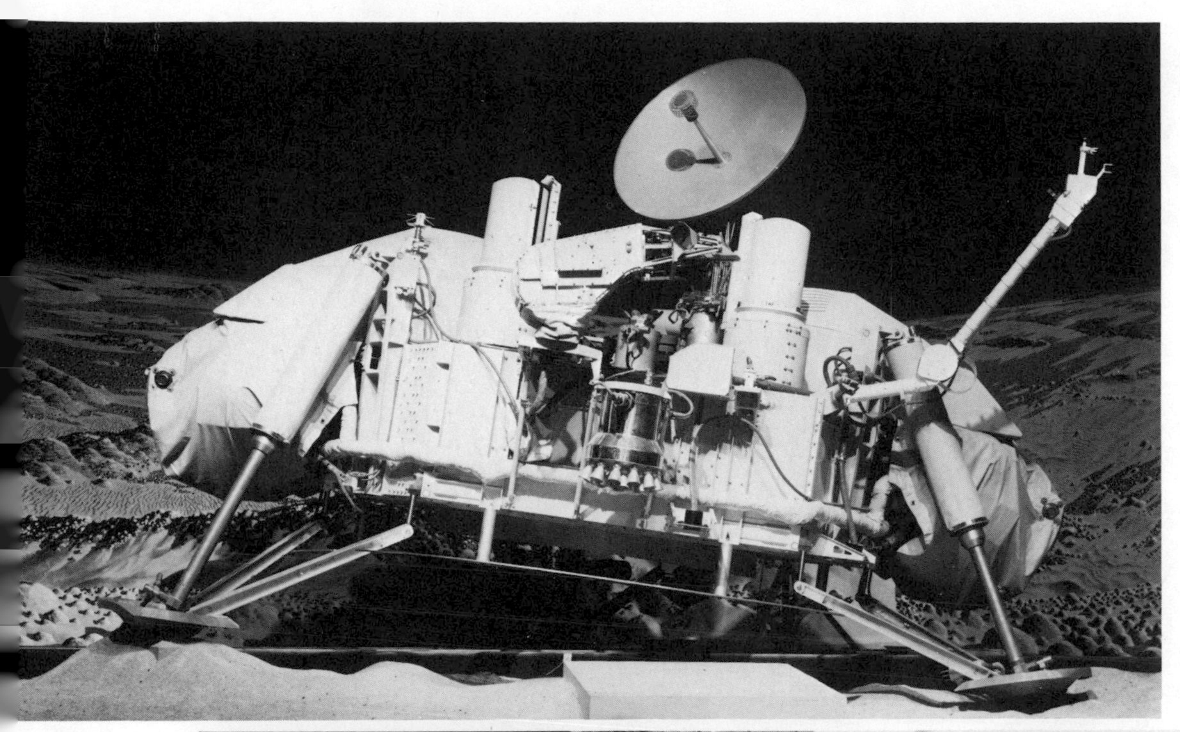

George is a medical robot that has several heart attacks every week to help train future doctors.

Remarkably enough, George always recovers, no matter how serious the illness is. The reason is that George is a robot. He is a very human-like robot; even his plastic skin looks and feels natural. But hidden inside his bed is a computer that is programmed to present a full range of symptoms for several different heart diseases.

George is one of a growing number of robots used to teach medical students. Other robots are being used to teach ophthalmology (diagnosis and treatment of eye disease), dentistry, and other fields of medicine.

DOMESTIC ANDROIDS

Domestic androids are robot servants. They are able to do household chores such as walking the dog, vacuuming the rug, taking out the trash, bringing in the mail, mixing and pouring drinks, and dozens of other jobs around the house. Some can hear and obey spoken commands, and some can even speak themselves.

Arok, made by Ben Skora, and Klatu, an invention of Tony Reichelt, are two of today's best-known domestic androids.

Arok is 6 feet 8 inches (2 m) tall, and weighs 275 pounds (125 kg). His body, head, and arms can bend and turn. His lips move as he speaks. He is strong enough to pick up weights of over 100 pounds (45 kg). Arok walks by means of wheels hidden in his legs. He can either go forward or backward at speeds up to 3 miles (5 km) per hour.

Arok contains fifteen electric motors that allow him to move thirty-six different ways. These motors can be directed either by an operator at a control panel or by a memory tape inside the robot.

Right now, the operator can only control Arok if he or she can see the robot. Skora, though, is planning to install a television camera in the robot's head, so the operator can control the robot from a distance. He is also building a more advanced, 3-foot (1-m) tall cylindrical robot. It will have an especially strong artificial intelligence.

Klatu is similar to Arok. He is also a domestic robot, even though he has been used mostly at meetings, conventions, and opening ceremonies of stores. Klatu is 5 feet 4 inches (1.7 m) tall and weighs 240 pounds (109 kg). He can answer questions, play games, and make his way through a crowd, all under the remote control of a human operator. He can bend his arms, elbows, and

JAROK

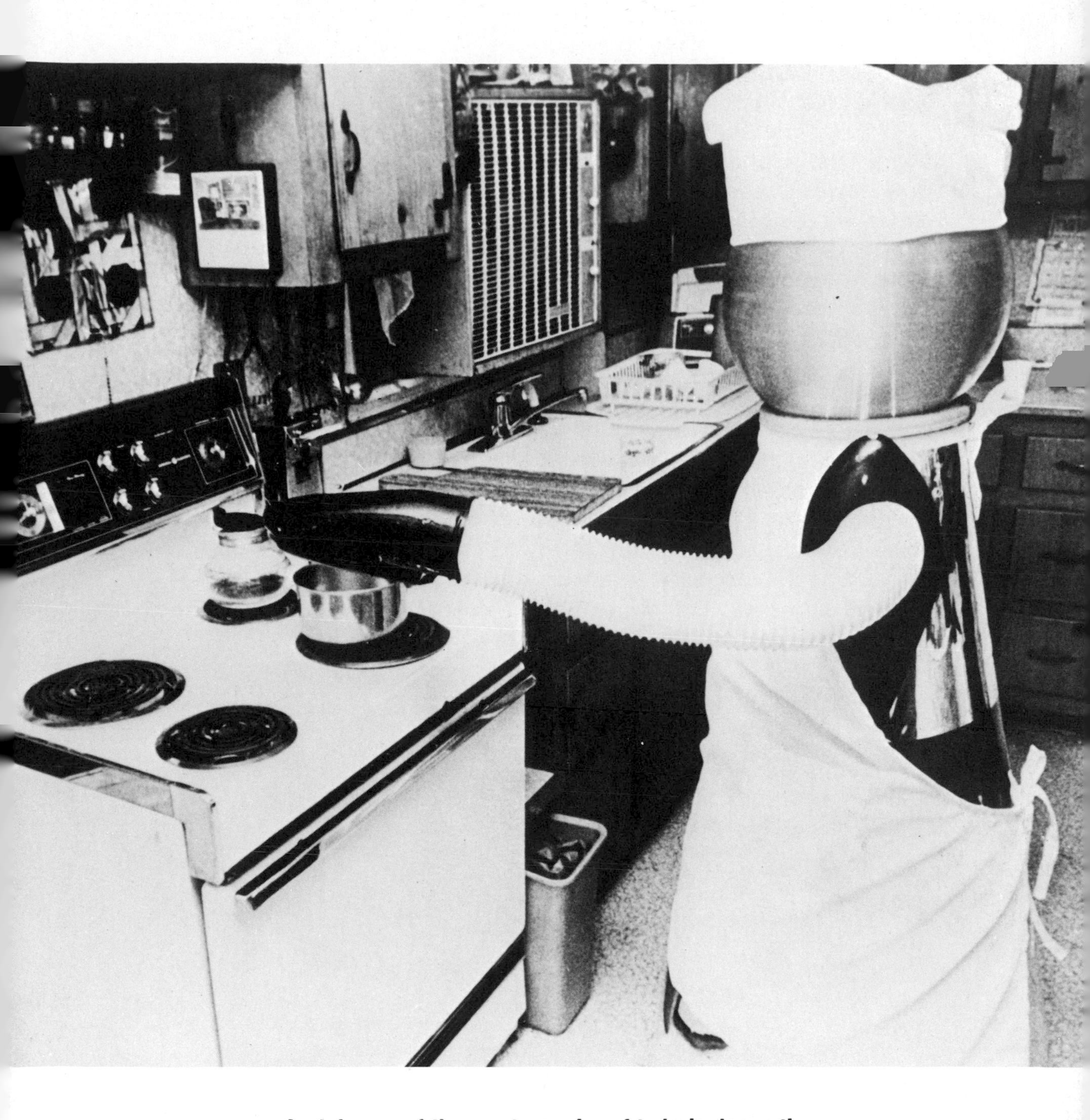

Arok is one of the most popular of today's domestic androids. He is shown at left vacuuming the carpet. Klatu, a well-known domestic android, wears an apron and a chef's hat as he cooks dinner.

hands, and when he speaks his translucent head emits flashes of light in time with the words.

Tony Reichelt, the creator of Klatu, is now planning Century I, a robot security guard that will be able to zip around empty stores or factories at speeds of up to 30 miles (48 km) an hour. This clever sleuth will be able to detect an intruder and give an electric shock or a blast of tear gas if one is found. Reichelt is also working on Sally, a robot that will go up and down the aisles of department stores announcing the sales and specials of the day.

For the future, Reichelt is contemplating the development of a second-generation domestic android. This new robot, when built, will be able to do much more than today's domestic robots. It will answer the doorbell, hang up coats, and announce the guests; serve meals; vacuum, polish, and dust; watch out for fire or burglars; act as a nurse by sounding an alarm if there is a change in a patient's condition; and recognize up to six members of a family by their voice patterns.

Until now, domestic androids have mostly been used for exhibition purposes. But there is little doubt that as they are improved and accepted, they will be put to work in many homes.

CHAPTER 5

ROBOTS OF MYTH AND MAGIC

ANCIENT TIMES

In the dark, quiet hours just before dawn, men, women, and children stream out of their houses. They head across the cool desert sands to the immense temple erected to pay homage to their dead ruler, Amenhotep III.

The people form a vast throng at the feet of the two colossal statues of Amenhotep that guard the temple entrance. As dawn breaks, the crowd grows silent. All eyes turn up to watch the head of one of the huge stone statues.

And then it happens—as it happened every dawn for hundreds of years in the ancient city of Thebes in Egypt. Just as the first rays of the sun strike the statue, the statue speaks. It makes sounds like the soft, lovely tones of a harp. Some think they hear a message of great wisdom from their departed ruler. Some claim they see the statue's stone lips part as it speaks. For many, this experience confirms their strong belief in life after death.

After a few minutes, the statue becomes silent. Then the people start back on the long walk to Thebes to begin their day's work.

Centuries later, when the Greeks conquered Egypt, they renamed the talking statue Memnon, the son of dawn. The Greeks also heard the statue speak at daybreak. But they said it was Memnon greeting his mother Eos, the goddess of dawn.

The statue of Amenhotep dates back over 3,000 years. Yet long before it was made, there were already other legends about statues that came to life. There were myths concerning the magic that could give lifeless stone or metal the ability to move and to speak.

In the epic poem *The Iliad,* Homer tells the story of Vulcan, the Roman god of fire. (Vulcan was called Hephaestus by the Greeks.)

According to Homer's account, Vulcan built twenty young maidens of gold. He fitted them with "minds of wisdom," and they moved about on "wheels of gold." He used them to serve in his dining hall. Homer said these early robots were "miraculous to see."

Vulcan's most miraculous robot was Talos, a huge bronze mechanical man. Vulcan created Talos for Minos, king of Crete, to protect the island from its enemies. According to legend, Talos made three complete circles around the coast every day. If he saw enemy ships nearing the island, he flung huge stones at them. If enemies landed on the island, Talos became red-hot, grasped the intruders in his powerful arms, and burned them to death.

Talos had one weakness, however, an opening in his foot. This opening was normally kept closed by a plug. One day a group of enemies landed on Crete. One of them pulled the plug, and the liquid that gave Talos life poured out. He collapsed and disappeared.

MIDDLE AGES AND RENAISSANCE

There is a story that the famous thirteenth-century scholar, Albertus Magnus, worked for thirty years to create a mechanical being that could move, talk, think, and perform a number of tasks.

The colossal statue of Amenhotep, or Memnon, spoke every dawn.

When finished, the robot could answer questions on religion and logic and could solve problems in arithmetic. The robot also guarded Magnus's rooms in Cologne, Germany. When someone knocked, the robot went to the door, opened it, greeted the visitor, and asked why the visitor had come.

One day, the great church leader Thomas Aquinas came to call on Magnus. Aquinas was offended at the sight of a machine in the shape and form of a man. Believing it to have something to do with the devil, he flung the robot into the fireplace, completely destroying it.

A tale about the renowned English friar and thinker, Roger Bacon, also in the thirteenth century, tells how he and another friar built out of brass a head that could speak. It seems that the two men wished to build a wall around Britain to protect it from its enemies. While they were deciding how the wall should be built, Bacon had a vision. He was told to construct a brass head. The head would then tell him how to build the wall.

It took Bacon and the other friar seven years to complete the lifelike brass head. For three weeks, the two churchmen watched the statue day and night, waiting for it to speak. It said nothing. Eventually they grew weary. They asked a servant to watch the head while they slept for a few hours. The servant had strict instructions to wake them at once if the head began to speak.

No sooner were they asleep than the lips parted and the head said, "Time is." These two words did not seem very important to the servant, and he let the two men sleep on. A half hour later the head spoke again. "Time was," it said. Again the servant decided not to wake his masters. After another half hour, the head spoke for the third time. "Time is past," it said. And so saying it collapsed and was never heard from again. Needless to say, the wall around Britain was never built.

THE GOLEM

In the middle of the sixteenth century, according to Jewish legend, Rabbi Loew of Prague, Czechoslovakia, created a robot to protect the Jews from their enemies. With the help of two assistants, Rabbi Loew shaped a 9-foot (2.7-m) tall man of clay. He called the figure Golem, which means something that is incomplete or not fully formed.

When the Golem was finished, one of the helpers circled the statue seven times from left to right while the Rabbi said some magical prayers. The Golem began to shine and glow as though he were afire.

Then the other helper circled the statue seven times from right to left as the Rabbi said other prayers. This time the glow went out, smoke arose from the figure, hair appeared on his head, and nails grew on his fingers.

Now the Rabbi himself went round the Golem seven times from left to right. Then he stopped and pronounced the secret name of God, whereupon the Golem received the spark of life and opened his eyes.

The legend goes on to tell how the Golem protected the Jews and killed those who would destroy the Jewish people. But the Golem did not stop there. He began killing innocent people, including Jews. Eventually, the Jewish people had to destroy the Golem.

In 1914, the first of two films, both called *The Golem,* was made in Germany. The story concerns an antique dealer who buys an old statue that he recognizes as the Golem. He places a magic charm around the Golem's neck, and the statue comes to life.

The Golem falls in love with the antique dealer's daughter, but she spurns his affection. The rejected robot goes beserk. He rampages through the streets, killing everyone and destroying everything that gets in his way. Finally, though, the antique dealer's daughter is able to snatch the charm from his neck. The Golem turns back to lifeless stone.

This 1914 film was so successful that Paul Wegener, who wrote and directed it and played the part of the Golem, decided to do a second version in 1920. This film, also called *The Golem,* had the subtitle *How He Came into the World.* It is a little closer to the old Jewish legend.

A young army officer comes to Rabbi Loew with a decree from Emperor Ludwig that all the Jews are to leave Prague within a month. The Rabbi models the clay figure of the Golem to help his people. He writes the magic word "Aemaer" on a slip of paper and places it in a charm around the Golem's neck. With that, the creature comes to life.

The Golem becomes a perfect servant for the Rabbi. He even goes to market with a shopping bag on his arm. In one famous

The Golem is a legendary robot that is said to have protected the Jews in Prague during the sixteenth century. This picture shows the Golem as he was portrayed in a 1920 film.

scene, the Golem is shown hacking away at something that cannot be seen on the screen. The audience is convinced he is committing some horrible murder. Instead, they discover he is chopping wood for the Rabbi's fireplace.

Meanwhile, the young officer falls in love with the Rabbi's daughter. One day the Rabbi finds the pair together. He orders the Golem to kill the officer. The Golem flings the young man to his death from a tower.

In the riots that follow, everyone flees the Golem. Only one little girl smiles at him. She offers him an apple. He accepts and picks her up in his arms to show his love.

The little girl begins to play with the charm around the Golem's neck. At one point she opens the charm and playfully removes the paper with the magic word. In an instant the Golem stiffens. He becomes a lifeless stone statue once again.

R.U.R.

Although mythical and magical robots go far back in history, it was not until 1920 that the word *robot* was coined. The word was used by Karel Capek, a Czechoslovakian author, in his play, *R.U.R.* The initials stand for *R*ossum's *U*niversal *R*obots. Capek derived the word *robot* from the Czech word *robota,* which means "worker" or "drudge."

Capek's play is set on an island in the not-too-distant future. In his laboratory the scientist Rossum finds the magic formula to create living robots. These intelligent, efficient, tireless robots are free of all human emotions. They are sold by the hundreds and thousands as workers and soldiers for the factories and armies of the world.

Helena Glory, president of the Humanitarian League, decides she will try to make life more pleasant for the robots. She convinces Dr. Gall, head of research in the robot-manufacturing plant, to change the formula slightly so that the new robots will have feelings.

The humanized robots stir up a worldwide revolt. They succeed in destroying all humans save Alquist, the person in charge of production at the R.U.R. factory. They hope, in vain, that Alquist will help them find the secret formula for making feeling robots, which was lost during the revolt.

It was in the 1920 play *R.U.R.*, that the word robot was first used. At the climax of the play, the robots revolt and kill all the humans except one.

Just when it seems that both humans and robots are going to become extinct, two robots, a male and a female who are more human than the other robots, fall in love. They go off together at the end of the play, like Adam and Eve, the implication being that the human race may begin again.

Capek's message was clear: Working in a factory or serving in the military turns people into living robots. In time, this will lead to the destruction of our civilization. The only hope for the future is to develop and stress the feeling, caring, emotional sides of ourselves.

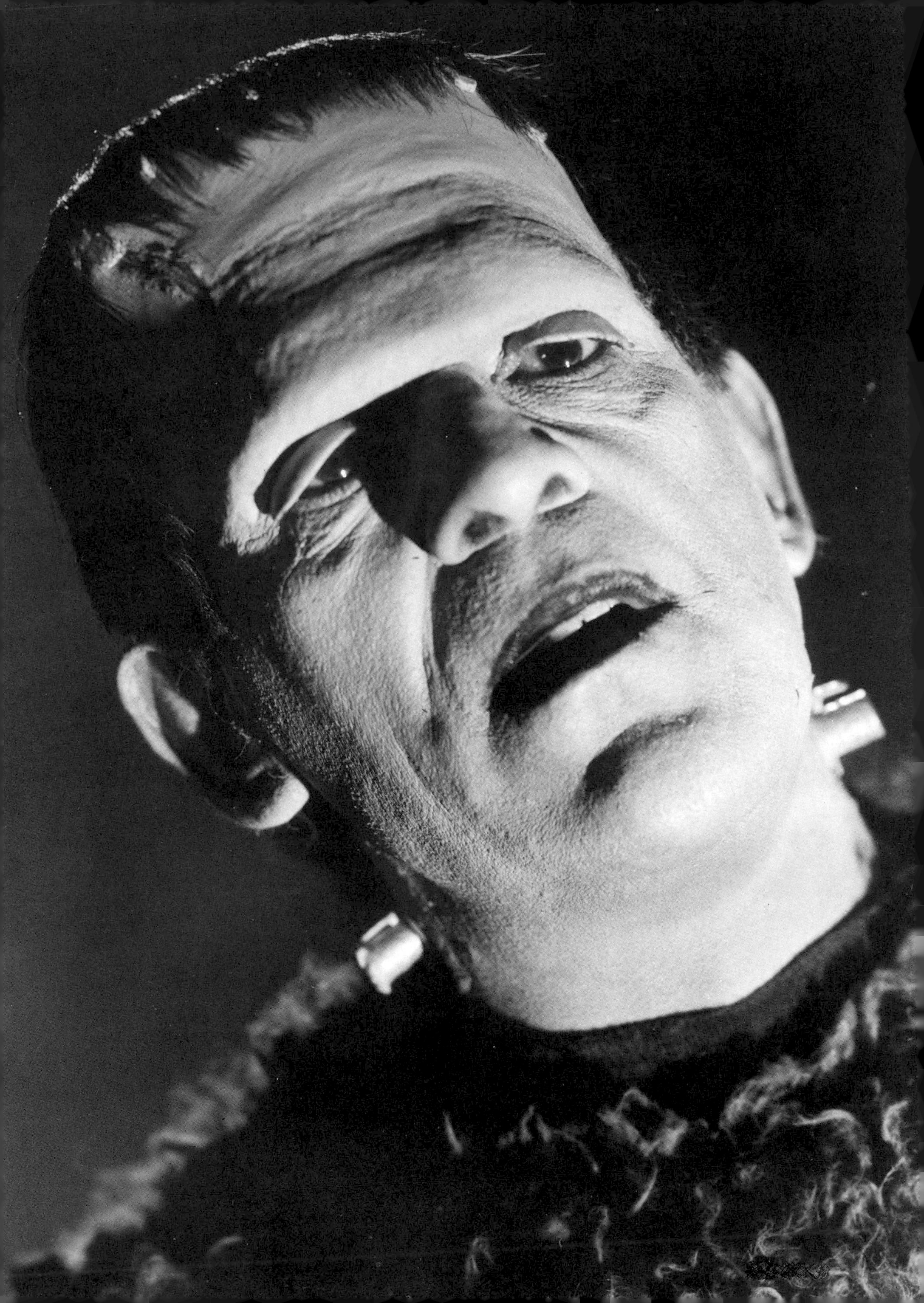

CHAPTER 6

HUMAN ROBOTS

FRANKENSTEIN

Mary Shelley's novel *Frankenstein* was written during a rainy summer holiday in Switzerland more than one hundred fifty years ago. To help pass the time, Mary, her husband the poet Percy Bysshe Shelley, and their friend Lord Byron, also a poet, each decided to write a ghost story.

At first, Mary found it hard to think of a dramatic and frightening plot. But then some ideas came to her. She thought over the endless debates about whether or not there was a "spark of life." And she recalled the fairly recent research on electricity, particularly Benjamin Franklin's famous experiments with lightning. (Some believe that Mary Shelley created the name Frankenstein from the name of the American scientist-inventor-statesman.)

The tale of horror that Mary Shelley published in 1818 was the first work of fiction to deal with the creation of a living robot from the corpses of humans. Since then there have been a host of films, books, and plays, all based, more or less, on the original Frankenstein novel.

Because Frankenstein's demon is unable to find friendship and love, he becomes a murderer and a creature of evil, as seen in this still from the film *Frankenstein.*

In the story, Victor Frankenstein, a brilliant but slightly mad scientist, learns the secret of life. He uses this information to create a monster, or demon. The demon is made up of parts of dead bodies that he collects from the morgue. The demon has no name. (Frankenstein is the name of the scientist, not the demon.)

Mary Shelley describes the demon as follows: "His yellow skin scarcely covered the work of muscles and arteries beneath; his hair was of a lustrous black, and flowing; his teeth of a pearly whiteness; but these luxuriances only formed a more horrid contrast with his watery eyes, that seemed almost of the same colour as the dun-white sockets in which they were set, his shrivelled complexion and straight black lips."

Frankenstein is disgusted by the demon's ugliness. Even though it is his own creation, he flees. The demon, left alone, sets out in search of love and companionship. He comes to a forest where he has a series of bad experiences. On every side he is spurned and rejected. Eventually he gives up his search for friendship and is overcome with anger and hatred. He vents his great rage in murder.

One day Frankenstein and the demon accidentally meet. The demon asks Frankenstein to create a companion for him, someone he can love and who will love him. Frankenstein agrees. Just before the new being is finished, however, Frankenstein becomes frightened that the demon and his mate will give birth to a race of monsters. He stops his work and cuts the creature into pieces.

When the demon discovers that Frankenstein has broken his promise, he becomes further enraged. He vows to avenge this cruel act. More murders follow. Finally Frankenstein is killed, and the demon flees into the icy Arctic wilderness.

Frankenstein established three ideas concerning robots: That robots will turn on their masters and even try to kill them (death awaits anyone who tries to play God by creating human life); that inert matter can be given life by means of electricity; and that robots are basically sympathetic characters who seek love and warmth and only become a menace when these necessities are denied them.

HUMAN ROBOTS IN THE MOVIES

The first film on the theme of the creation of a human robot was *Frankenstein,* made by Thomas Alva Edison in 1910. Unfortunately, however, all prints of this film have been lost.

The great German film made in 1919, *The Cabinet of Dr. Caligari,* also concerns the creation of a human robot. This robot, though, is not made from parts of corpses. Rather, he is a living man who is placed under a spell that robs him of his willpower, his emotions, and his ability to think.

The central character of the film is Dr. Caligari, a showman and fortune-teller, who appears at fairs and carnivals with his assistant, Cesare. Dr. Caligari has hypnotized Cesare and the young man is in his power. He keeps Cesare in a coffin-like cabinet as they travel from fair to fair.

The doctor boasts to his audience that Cesare knows all secrets. At one village fair a young man, Alan, asks Cesare, "How long shall I live?" Cesare replies, "Until dawn." To make the prophecy come true, Caligari orders Cesare to kill Alan.

Alan's friend, Francis, is determined to find the murderer. He traces the crime to Caligari, follows him, and discovers that Caligari is really the director of the local mental hospital. In a final twist of the plot, Francis is found to be a patient at the hospital. The whole story turns out to be one of Francis's nightmares.

The great film classic, *Frankenstein,* was made in 1931. Boris Karloff became a star after his outstanding performance as the monster.

At the beginning of the film, Dr. Frankenstein and his misshapen assistant, Fritz, are collecting corpses from a cemetery. Later, Dr. Frankenstein instructs Fritz to steal a preserved brain from the medical school. By accident Fritz steals the brain of a criminal.

In his laboratory, Dr. Frankenstein creates a person from parts of various corpses and the brain of the criminal. The figure, covered with bandages, lies motionless on a laboratory table. All it needs is the electrical power of a lightning storm to give it life.

At last a storm comes. Bolts of lightning flash across the tower that looms above the laboratory. As the storm reaches its climax, Dr. Frankenstein gives a signal. Fritz turns a handle that slowly raises the body to the top of the tower. Powerful bolts of lightning strike the lifeless form. The electricity surges and crackles.

Then, as the storm subsides, the assistant lowers the wrapped body. Frankenstein stares at the hand that hangs over the side of the table. As the hand slowly begins to move, he cries out, "It's alive—it's alive!"

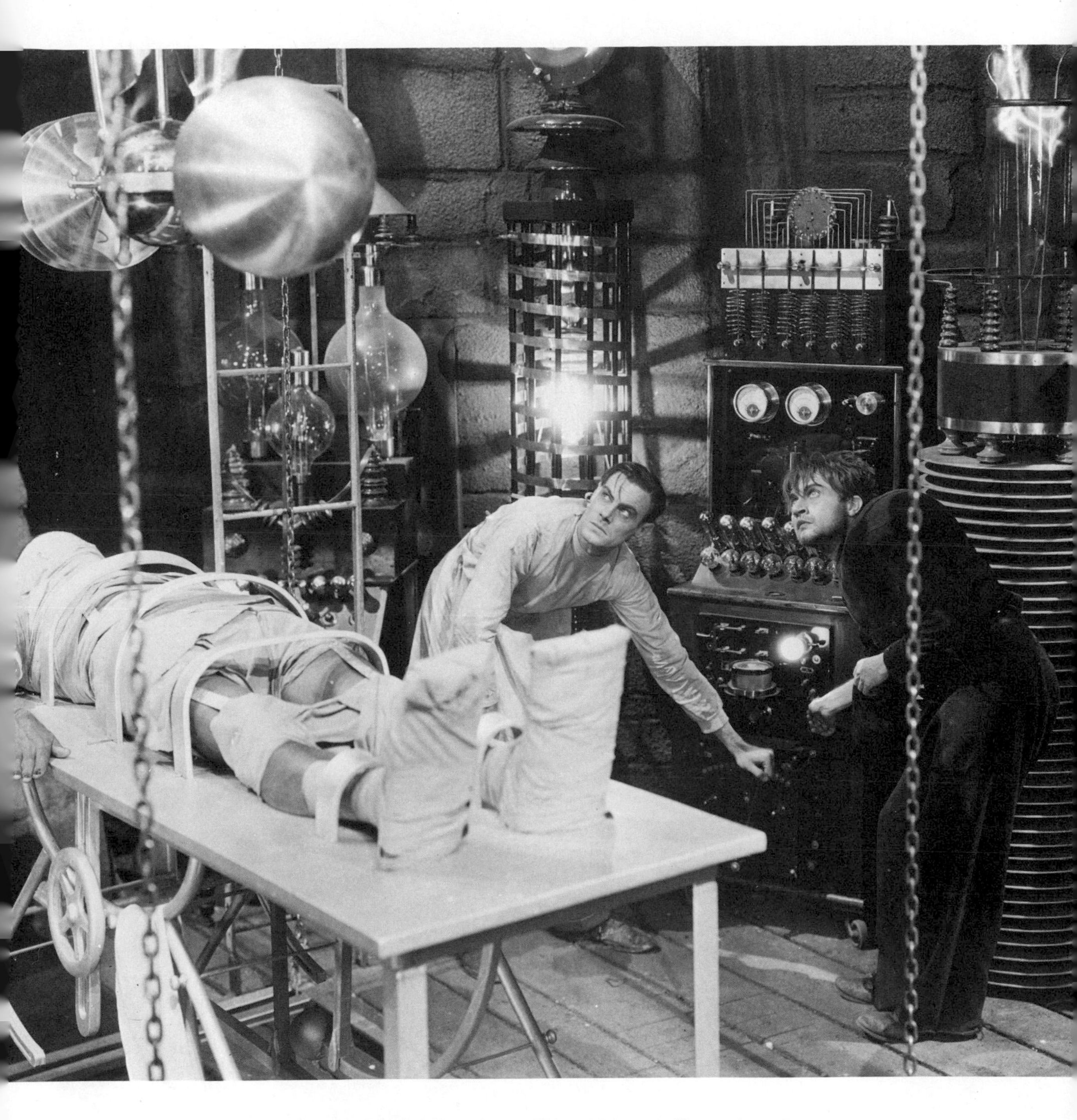

Dr. Frankenstein and his assistant have collected parts of corpses that they have assembled into a body. In this scene from *Frankenstein,* they are preparing to raise the body to receive a life-giving bolt of lightning.

This is the famous scene in which the monster is playing with Maria. Later he kills her, not realizing what he is doing. This scene was eventually cut from the film.

The monster is a frightening sight. His head has a square shape and his forehead is unusually high. Scar lines show where Dr. Frankenstein had stitched the body parts together. Metal plugs extend on either side of the neck. Thick and heavy eyelids hang down over partially closed eyes.

The monster's body is large and clumsy. When later he walks, it is with a stiff gait. His black fingernails are a constant reminder that the monster was created from corpses.

For all its horrible features, the monster still somehow looks pathetic and sorrowful. It seems to want to be friendly and expresses a need to give and receive love. (Most people agree that this side of the monster's personality contributed to the immense success of the film.)

But love and friendship are not what the doctor planned for his creation. The monster is whipped and chained and subjected to scientific study and examination. This treatment soon changes the monster into an angry and destructive creature.

In one famous scene, after he escapes his captors, the monster comes upon a young girl, Maria, playing by the side of a lake. He joins her in picking daisies, throwing them into the water, and watching them float away.

After a while there are no more daisies to float on the water. So the monster picks up Maria and gently tosses *her* into the lake. The idea was to show that, in his simple way, the monster regarded Maria as a flower that would float along with the other flowers. Of course the girl drowns.

Audiences were confused by the monster's action. Some interpreted it as very cruel. Others thought it revealed the monster's lack of understanding. Still others just laughed, because they could not fathom its meaning.

So the part where the girl is tossed into the lake was eventually cut from the film, causing the story to jump from the two playing together to the father carrying the drowned child's mud-caked body back to the village. The new version makes it appear that the monster set out to kill the child.

Citizens of the village call for the capture and execution of the monster. With flaming torches and killer dogs, they set out to find him. Finally he is chased into a large wooden windmill, where he finds and kills Dr. Frankenstein. The villagers set the mill on fire, and the monster is destroyed.

After the great success of *Frankenstein,* a sequel was made in 1935, *Bride of Frankenstein,* which brought both Dr. Frankenstein and the monster back to life. Of all the film versions of Mary Shelley's story, this one treats the monster most sympathetically.

After escaping from the burning mill, the monster finds shelter in the hut of a blind hermit. Here at last he is regarded as a friend rather than an object of fear and loathing. The hermit gives him bread and wine, plays the violin for him, and teaches him to say a few words.

The happy time ends, however, when a hunter enters the hermit's hut and drives the monster out. The monster is once again filled with hatred and violent feelings.

But Dr. Frankenstein has decided to create a female robot as a mate for the monster. Using the same means he used to create the monster, he brings a woman to life.

The monster and his bride are brought together. In a tender gesture of love, the monster stretches out his arms toward the newly created woman. But she takes one look at his face and shrieks in dismay.

"She hate me, like others," the monster says, speaking slowly and haltingly. A single tear rolls down his cheek. Stiffly his hand reaches for a lever set in the wall of the lab.

"Don't touch that lever—you'll blow us all to atoms!" shouts Dr. Frankenstein's assistant.

As the monster pulls the fateful lever, he utters the famous last line of *Bride of Frankenstein,* "We belong dead." There is a huge explosion, and the monster and his bride are destroyed in the flames.

Despite the violence of the explosion, Hollywood was able to bring the monster back for a number of sequels. *Son of Frankenstein* followed first in 1939; *The Ghost of Frankenstein* came next in 1942. *The Curse of Frankenstein, The Revenge of Frankenstein,* and *I Was a Teenage Frankenstein* were some of the less artistic efforts to cash in on the fame of the first two Frankenstein movies.

In 1941, the film *Man-Made Monster* carried on some of the ideas from *Bride of Frankenstein.* Dr. Rigas, a mad, evil scientist learns of Dynamo Dan, a carnival performer who does various tricks with electricity. Rigas brings Dan to his lab and feeds his body with greater and greater amounts of electricity.

"She hate me, like others," the monster says, when the bride that Frankenstein has created turns away from him in the film, *Bride of Frankenstein.*

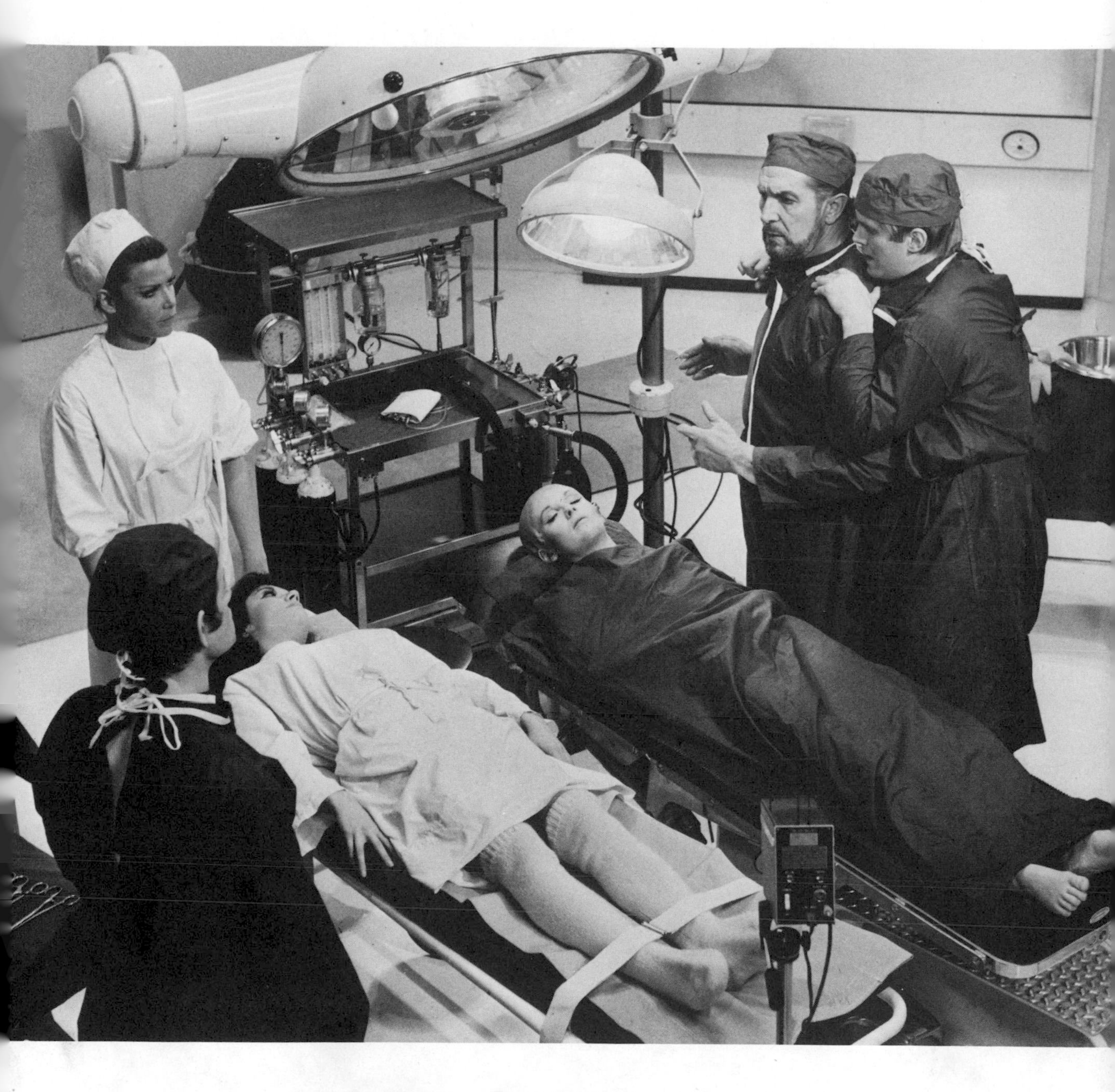

Left: Dr. Rigas has fitted Dan with a rubber suit to contain the powerful electrical current flowing through his body. Above: in *Scream and Scream Again,* Dr. Browning collects body parts to build perfect human robots.

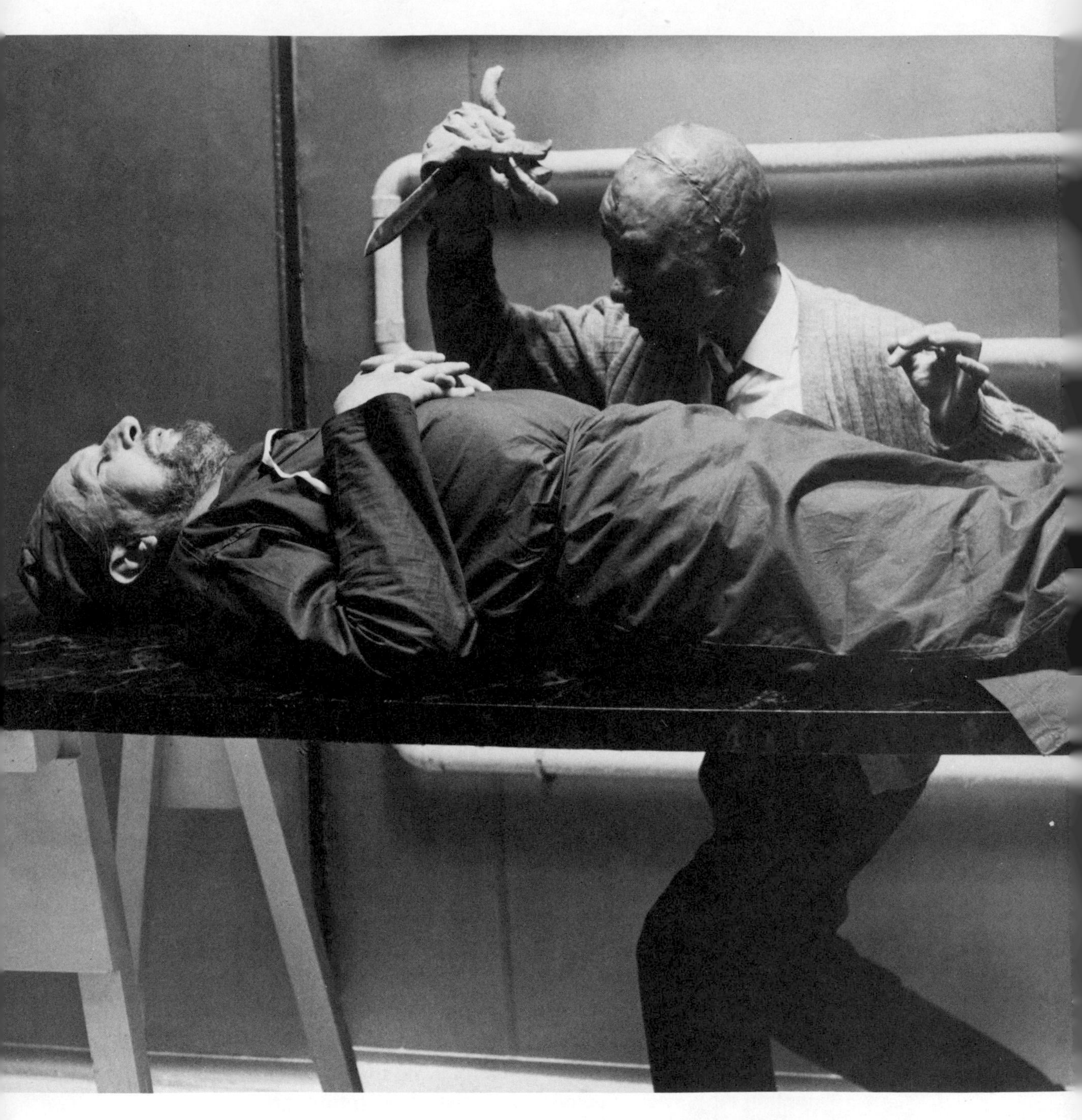

One of the most frightening scenes of *Scream and Scream Again* comes when one of the human robots tries to kill Dr. Browning.

The Doctor builds up such a tremendous charge of electricity in Dan's body that it robs Dan of his will and turns him into a robot. When someone on the outside learns of Dr. Rigas's experiments and threatens to expose him, the doctor orders Dan to kill the person.

Dan is caught and sentenced to die in the electric chair. But the powerful electrical current does not harm him. He returns to Dr. Rigas, who fits him with a rubber suit to contain the electricity.

When Dan finds out that Dr. Rigas is planning to create a female electrical robot out of a woman, he tries to stop him. In a terrifying shower of sparks he kills Dr. Rigas, frees the woman, and carries her off into the night. He stumbles in the dark, though, and tears his rubber suit on a barbed wire fence. His electricity leaks out and he is burned to an ash. The woman escapes to safety.

The 1970 horror film, *Scream and Scream Again,* is another, more recent, retelling of the Frankenstein story. This time the evil genius, Dr. Browning, murders in order to collect different body parts. His ambition is to build perfect human robots. The climax to the complicated plot occurs when Dr. Browning himself is revealed to be a robot. Demolished in a bath of acid, he joins the many other fictional characters who met similar fates because they tried to create life in the laboratory.

CHAPTER 7

SCIENCE FICTION ROBOTS

Most of the fictional robots that have appeared in film, on television, and in books and comics were based on real scientific ideas. They are usually known as science fiction robots.

A staggering number of science fiction robots have been created over the years. Most have already been forgotten. But others have captured people's interest and continue to be popular. Here is a sample of some of the most enduring of these robots.

EVIL ROBOTS

Maria, one of the first science fiction film robots, appeared in Fritz Lang's 1926 film classic, *Metropolis.* A real, flesh-and-blood human named Maria is leading a worker's revolt in the modern city of Metropolis in the year 2000. The ruler of the city orders Dr. Rotwang, a scientist, to create a robot that will look exactly like the real Maria. His plan is to use the robot Maria to cause a riot, which in turn will cause a flood. The waters from the flood will swamp the workers' underground homes and kill the workers and their children. Then the dictator will make robots to take the workers' places.

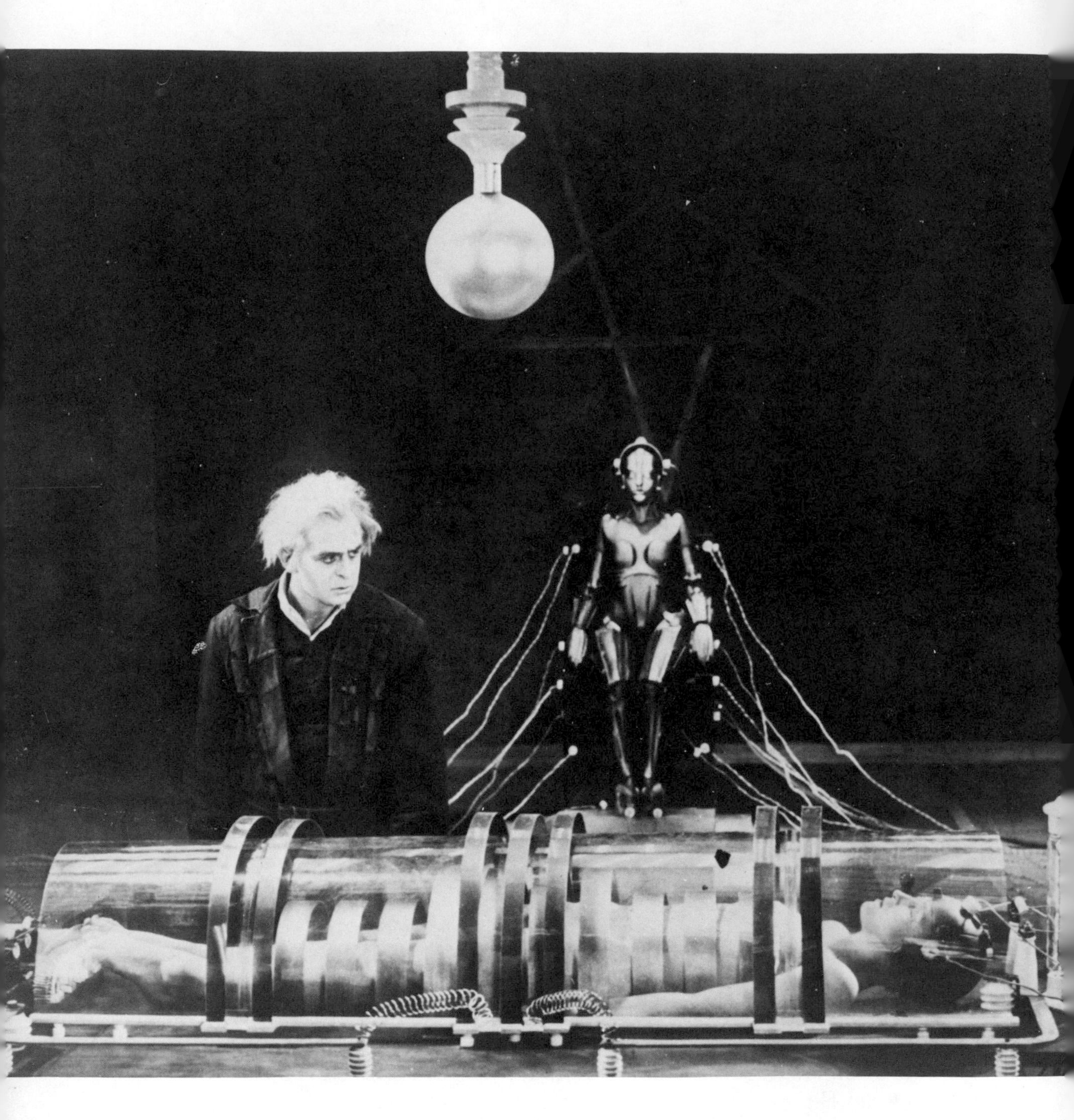

The evil scientist Rotwang
gives the robot Maria life.

Rotwang builds a metal robot that he connects by wires to the real Maria, who is his prisoner. The metal robot comes to life. It looks and acts exactly like the human.

The robot Maria almost carries out the plan. But at the very last moment, the real Maria escapes from Rotwang's lab and saves the children. The workers decide to burn the robot Maria at the stake. In the flames, she loses her resemblance to the real Maria and becomes nothing more than a mechanical robot. The real Maria persuades the ruler to grant the workers some of their demands. In a happy ending, the ruler and the leader of the workers shake hands.

A robot named Annihilaton appears in the film *Flash Gordon Conquers the Universe,* made in 1940. The robot is sent from the planet Mongo by Ming the Merciless to prevent Flash Gordon and his girlfriend Dale Arden from finding the cure for the Purple Death plague, which has been sent by Ming to destroy the people of earth.

Annihilaton returns to Mongo with Dale as his prisoner. But in a bold move Flash rescues her. Then he crashes an empty spaceship into Ming's palace, killing the ruler and destroying his wicked empire.

Flash Gordon Conquers the Universe is a serial—a series of short films all telling one story. Serials were quite popular during the 1930s and 1940s. They were usually shown on Saturday afternoons in local cinemas. Each film told only one episode in the hero's exciting life, and each one left off at a climactic moment, so that the audience would come back the following week to see the next episode.

Some of the most fanciful and imaginative—and evil—of all robots are found in the comics. In the story of "Magnus, Robot Fighter, 4000 A.D.," Magnus and his beautiful girlfriend Leeja are called on to fight the wicked robot, Bunda the Great, who is trying to enslave the world. By using his great strength and brilliant mind, Magnus triumphs over Bunda. Magnus's victory symbolizes the mastery of humans over machines.

Each issue of *Where Monsters Dwell* features a robot adventure. One tells of the landing on earth of the giant robot, Orogo, in his immense spaceship. With his single eye, Orogo hypnotizes large numbers of people. Then he orders them into his spaceship, for transportation back to his own galaxy as slaves.

Above: in the serial *Flash Gordon Conquers the Universe,* the robot Annihilaton kidnaps the heroine, Dale Arden. Many fictional robots appear in comic books. "Magnus, Robot Fighter," (right) fights evil science fiction robots.

GOLD KEY ROBOT FIGHTER 12¢

10046-711
NOVEMBER

MAGNUS ROBOT FIGHTER

4000 A.D.

MAGNUS COMBATS THE CLUTCHING CLAW OF
BUNDA THE GREAT!

WHERE MONSTERS DWELL
MARVEL COMICS GROUP
APPROVED BY THE COMICS CODE AUTHORITY
20¢
26
JAN
02486
WHERE MONSTERS DWELL
WHAT WAS THE DREAD SECRET OF...
"THE THING CALLED METALLO!"
THERE IS NO PLACE TO RUN, HUMANS! METALLO IS HERE!

It seems that nothing can stop the mighty Orogo from casting his hypnotic spell. Then one old man challenges the robot. He says that Orogo cannot hypnotize him.

Orogo stares down at the old man, but he is not hypnotized. Orogo increases the power of his look. Still the man is unaffected. Orogo makes his beam stronger and stronger, but with no success. Finally Orogo exceeds his limit. With a thunderous crash, the robot collapses to the ground.

Later, when asked how he was able to withstand Orogo's most powerful look, the old man quietly says, "I saw nothing. You see, gentlemen, I am totally blind!"

Another issue describes "The Thing Called Metallo!!" An escaped killer, Mike Fallon, convinces the army to let him test Metallo, a hollow robot controlled from within by a man. Metallo is built to withstand any attack, even an atomic bomb explosion.

Once Fallon has learned how to operate the robot, he uses Metallo to pursue a life of crime. Nothing and nobody can stop him as he kills and robs, until he falls ill. The doctors tell him that the only treatment for his illness is radiation. Fallon is faced with a dilemma. If he stays within Metallo, the radiation will not reach him, and he will die of the illness. If he steps out of Metallo, the police will capture him and he will die in prison. Fallon decides to remain inside and goes off to await death.

GOOD ROBOTS

Not all fictional robots are evil. Many of the recent ones are "good" robots. A large number of these were created under the influence of Isaac Asimov, one of the best-known science and science fiction writers.

Metallo, in *"Where Monsters Dwell,"* was really a drone robot. He was controlled by a man from within.

Back in 1941, in his story *Caves of Steel,* Asimov stated the Three Laws of Robotics:

1. A robot may not injure a human being or, through inaction, allow a human being to come to harm.
2. A robot must obey the orders given it by human beings except where such orders would conflict with the First Law.
3. A robot must protect its own existence as long as such protection does not conflict with the First or Second Law.

Asimov's most famous robot book is *I, Robot.* The nine stories in the book all deal with robots that are made by the world's main robot manufacturer, U.S. Robot and Mechanical Men, Inc. Each story is about the conflicts that arise between humans and robots as robots are used more widely.

Probably the first of the good robot films was the 1951 hit *The Day the Earth Stood Still.* The robot in this film, Gort, and his master, Klaatu, arrive on a spaceship from another planet. They come to warn humans to stop making atomic bombs. They tell them that more advanced beings on other planets are afraid that earth people are going to blow up the entire universe. To show his power and determination to turn people from their destructive ways, Klaatu shuts off all the world's power sources, and the earth almost literally stands still.

Gort, the robot, has some remarkable powers, too. He brings Klaatu back to life after the alien has been killed by frightened humans. And he holds the entire U.S. army at bay by melting their guns and tanks with rays from his eyes.

After delivering their warning, Gort and Klaatu depart for home, hopeful that their message will be heeded.

The robot Gort uses his great power to bring his master back to life in the film *The Day the Earth Stood Still.*

The criminals in *Tobor the Great* use a computer to gain control of Tobor. The robot is then reprogrammed to attack the boy who built him.

Robby the Robot can cook, speak 188 languages, drive a dunemobile, and pilot a spaceship, as seen in this still from *Forbidden Planet*.

Tobor the Great (1954) is about a good robot (Tobor is robot spelled backwards) with emotions, built by an inventor and the inventor's grandson. Some criminals hear about Tobor and plan to kidnap the two who built it, in order to get control of the robot. But Tobor thwarts the efforts of the kidnappers and saves the scientist and the boy.

The would-be kidnappers then use a computer to gain control over Tobor. Under the computer's wicked power, the robot attacks the boy. Finally, though, the good side of the robot triumphs, and Tobor again serves the inventor and his grandson.

Forbidden Planet (1956) is set on the planet Altair IV in the year 2200. The planet is ruled by Dr. Morbius, who has created Robby the Robot, a devoted servant and companion. Robby prepares and serves the meals, drives the dunemobile all over the planet, and can translate 188 different languages.

A space patrol from earth lands on Altair, looking for a previous expedition that had never returned to earth. They learn that all the members of the earlier expedition had been killed by the dread Monster of the Id.

Soon the Monster appears and breaks into Morbius's home. Robby, who has always protected Morbius, is unable to save him now, and the doctor is killed by the Monster. The spacemen manage to escape, along with Morbius's daughter, just before the planet explodes, destroying both Robby and the Monster.

In *Forbidden Planet,* the Monster of the Id seems to represent the evil, antisocial, angry part of Dr. Morbius's personality. According to Sigmund Freud, the id is that part of the personality concerned with satisfying its own needs and desires without regard for others. Robby is unable to protect his master because the Monster is really part of Morbius's unconscious. The film makes the point that even scientific genius, as seen in Dr. Morbius's creation of Robby, can be destroyed by the unconscious evil drives present in each of us.

Robby the Robot was one of the most popular of all fictional robots. After appearing in *Forbidden Planet,* he was used in the 1957 film, *Invisible Boy.* Then, from 1965 to 1968, he played a major role in the television series, *Lost in Space.*

Lost in Space is a modern version of the nineteenth-century story *Swiss Family Robinson,* which is about the adventures of a family shipwrecked on a desert island. The family in the television series, also named Robinson, is selected to fly a robot-controlled rocket ship to the star Alpha Sentori. But an agent of an enemy government, Colonel Zachary Smith, sneaks on board and instructs the robot to destroy the ship. By accident, Colonel Smith is trapped inside the spaceship as it blasts off. He changes the robot's instructions, but not before it damages the ship, which crash-lands on an unknown planet.

Each episode deals with the Robinson family's attempts, with the help of the robot, first to survive and then to repair the spaceship for the trip to Alpha Sentori. Working against them is Colonel Smith, who frustrates their plans and tries to arrange for his own return to earth.

Robby the Robot, after the *Lost in Space* series ended, went into retirement. He is now on exhibit at the Planes and Cars of the Stars Museum in Buena Vista, California.

Robots appeared from time to time in the most popular science fiction television series of all, *Star Trek.* In one episode, called "What Are Little Girls Made Of?" the evil Dr. Korby lands on the planet Exo II. There he finds a tall, thin, hairless android named Ruk. Ruk is a firm believer in logical, step-by-step thinking.

Dr. Korby uses Ruk to start a race of super-robots that will conquer the world. Kirk, captain of the starship *Enterprise* and a fighter for good and decency, wants to stop Dr. Korby. He manages to convince Ruk that Dr. Korby's plans are illogical. This makes Ruk turn on Korby. Korby then destroys Ruk with a beam from his phaser gun. Thus ends Korby's dream of leading a robot army to victory.

Harry Mudd's robots in the "I, Mudd" episode of *Star Trek* are also highly logical and orderly in their thinking. They zealously serve the man who built them, King Mudd. So devoted are they, in fact, that they will not allow Mudd to leave the planet and return to earth.

When the evil Colonel Smith gains control over Robby the Robot in the television series *Lost in Space,* all sorts of adventures start to happen.

Star Trek was one of the most popular science fiction television series of all time. One episode featured Ruk, a tall, hairless android (left). In another episode of *Star Trek,* Harry Mudd builds a number of beautiful female robots (above) to lure the crew of the starship *Enterprise* to stay on his planet.

The robots Huey and Dewey in the film *Silent Running* are built by a scientist who fears that all the plants on earth will be destroyed by nuclear pollution.

The king brings the starship *Enterprise* to the planet Mudd by force. He uses a number of very attractive female robots to lure the crew of the starship into staying. But while Mudd is planning his escape, the crew of the *Enterprise* discovers his intentions. Their only hope is to get free of Mudd's robots. They plot an illogical scheme that confuses the robots and shorts out their circuits. The crew escapes. Mudd is left behind, still a prisoner of his robots.

The good robots in *Silent Running* (1972), Huey and Dewey, are comedians. They are the creations of a mad scientist who fears the destruction of all the plants on earth by nuclear pollution.

The scientist sets out in a rocket with Huey and Dewey to find another planet where he can grow plants brought from earth. One of the greatest robot comedy scenes ever filmed takes place in the rocket when the scientist and the robots play a game of cards.

The robots in *Westworld,* a science fiction film made in 1973, may be the most unusual robots of all. These robots are programmed to help the guests visiting a plush, expensive resort to realize their wildest dreams. The guests act out their fantasies in an old western town where they are the good guys and the robots are the bad guys —and the good guys always win. At the end, though, the robots revolt. They are no longer willing to die for the amusement of others. The struggle between robots and humans leads to the destruction of the robots and the death of many of the guests. *Futureworld,* the 1976 sequel to *Westworld,* continued the adventures of the human guests at a resort run by robots.

R2D2 and C3P0

But of all the fictional robots, none have so completely captured the public's imagination and affection as have the two robot heroes of the 1977 hit film, *Star Wars.* Originally written into the script for comic relief, the two robots, R2D2 and C3P0, emerged as major stars of the film.

C3P0, or See Threepio as his name is sometimes written, is a true android. Very polite, C3P0 acts and talks like a British butler. He is just under 6 feet (2 m) tall and has the shape and walking gait of a human. His skin is highly polished gold metal. A translator, C3P0 understands and speaks over 1,000 galactic languages. He can speak to people (organics, as they are called in the film), to robots

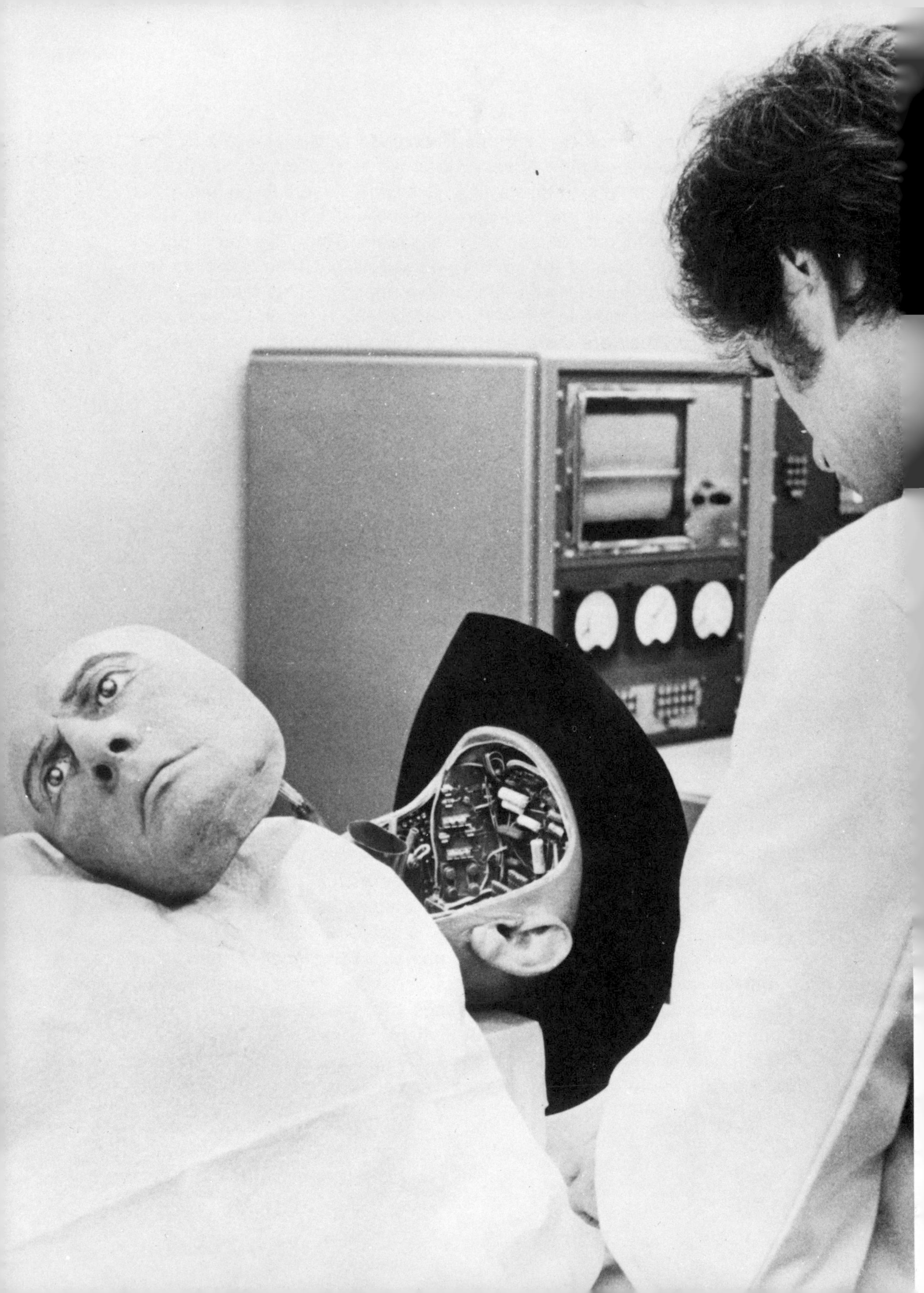

At the end of each day the robots in *Westworld*'s plush resort are readied for the next day's activities. Eventually, however, the robots in *Westworld* revolt and, as seen here, attack the guests.

(mechanicals), and to computers as well. He feels emotions and worries a great deal.

Threepio's electronic and mechanical features are based on very advanced robotics. His eyes are photoreceptors, which are sophisticated photocells that change varying light conditions into varying currents of electricity. His nose is an olfactory sensor at the base of his neck. A "vocabulator" produces speech. Microphones serve as ears to change sound vibrations into electricity. A band that looks like metal ribbon lies across the robot's head. This band is a microwave sensor and emitter that receives and broadcasts radio and other waves.

On Threepio's upper arms are energy controls. These receive energy from outside sources and change it into a form that allows the robot to move, think, sense, talk, and so on. The actual movements are made by servomotors, or self-governing motors. Set in the right side of Threepio's chest is a restraining bolt, a device that keeps the robot faithful to his organic owner.

R2D2, or Artoo Detoo, is the short, stubby partner of Threepio. Some say that his name is a pun on the letters R and D, the abbreviation for Research and Development, another name for scientific experimentation.

Artoo is only able to communicate with other robots and computers. The beeps, hoots, and whistles he produces form a highly condensed and compressed language that can only be understood by an information-handling machine.

Artoo is often called "a vacuum cleaner on legs." Only 4 feet (1.25 m) tall, he moves on two legs, or three when going over rough terrain. The two long legs on the side of his cylindrical body contain power cells that supply Artoo with energy. The additional short leg extends down from the middle of his body.

R2D2 and C3PO, the most famous of all fictional robots, appeared first in the film *Star Wars.*

Artoo is a utility robot, used to maintain and repair starships. Often he is used for sending messages. He is fed information that remains in his memory until it is no longer needed.

On top of the cylinder that makes up Artoo's body is a movable hemisphere, called the sensory input head. This area contains a single eye that sees by radar, rather than by photoelectric means. A knob on the top of the robot's head receives radio and electromagnetic transmissions.

The knob on the front of Artoo's head is an infrared receptor. It can sense sources of heat even when the source is far away. Artoo can use a laser beam to focus a 3-D hologram in space. Two panels on one side of the hologram projector connect the robot with computers for the exchange of information. Other panels on his body receive and give information to computers on other mechanicals. A grill-like movable panel in the middle of Artoo's body receives sound signals from outside. Next to this panel is the robot's restraining bolt.

Star Wars is set in a distant galaxy long ago. It concerns the efforts of the lovely Princess Leia Organa, the handsome young Luke Skywalker, Threepio, Artoo, and members of the Rebel Alliance, to bring freedom and democracy back to the galaxy. To do this they need to overthrow the heartless Grand Moff Tarkin and the Imperial Forces of the Galactic Empire.

Princess Leia and the two robots are in a spaceship that is captured by one of Tarkin's Imperial Cruisers. While Imperial Storm Troopers board the ship, Leia feeds Artoo plans on how to attack and destroy Death Star, headquarters of the Imperial Forces.

Princess Leia is captured, but the two robots manage to escape to the planet Tatooine. Here they are taken prisoner by the ugly Jawas and thrown into a sort of wandering junkyard for robots.

Eventually the Jawas sell the robots to Luke Skywalker's uncle, who plans to use them as workers on his farm. As Luke starts to repair and clean the robots, Artoo projects a hologram from Princess Leia, asking them to find Obi-Wan Kenobi, the only man Leia believes is capable of helping the rebel cause.

Luke, Threepio, and Artoo track down Obi-Wan. Before long, they are all on their way to Death Star in a spaceship piloted by Hans Solo and an ape-like creature, a Wookie named Chewbacca. A powerful tracking beam from Death Star, though, captures the spaceship and pulls it inside.

Unseen by the Imperial Forces, Artoo plugs into Death Star's control computer. He learns that Princess Leia is a prisoner there. Luke and the other rebels find Leia and free her from her cell. After averting several brushes with death, they escape to the rebel base on the planet Yavin.

Artoo now reveals the plans for the attack on Death Star. Luke jumps into his Incom skyhopper to lead the rebel forces in an attack against the Imperial Army. Artoo is plugged into the Incom's guidance system to guide Luke on his dangerous mission.

In a most exciting climax Luke, with Artoo's help, and Hans Solo fight off the Imperial Cruisers and blow up Death Star. Luke, Hans Solo, and the robots are all rewarded by Princess Leia for having freed the galaxy from the tyranny of the Galactic Empire.

Because of the great success of *Star Wars,* a television series called *Battlestar Galactica,* which closely paralleled the theme and special effects of *Star Wars,* was run. A sequel to *Star Wars,* entitled *The Empire Strikes Back,* was also soon put into production.

Artoo, Threepio, and other robots of fiction entertain and amuse us with their adventures in the world of make-believe. But they also suggest new directions and new ideas to scientists currently working in the field of robotics. Many features of today's fictional robots will be incorporated in the real robots of tomorrow.

At the same time, progress in the science of robots is influencing the creators of fictional robots. With each advance and improvement in real robots, writers and filmmakers are able to project their imaginations that much further into the future.

We enjoy reading about fictional robots and seeing them in movies and on television. We also benefit, directly or indirectly, from the robots of fact that are at work making our lives easier in the real world. Robots of fact and fiction are important to us now. They will surely be even more important in the days to come.

SUGGESTED READINGS

Asimov, Isaac. *I, Robot.* N.Y.: Fawcett, 1950.

Beck, Calvin Thomas. *Heroes of the Horrors.* N.Y.: Collier, 1975.

Berger, Melvin. *Bionics.* N.Y.: Franklin Watts, 1978.

Capek, Karel. *R.U.R.* N.Y.: Oxford, 1961.

Florescu, Radu. *In Search of Frankenstein.* Boston: N.Y. Graphic Society, 1975.

Geduld, Harry M. and Ronald Gottesman. *Robots, Robots, Robots.* Boston: N.Y. Graphic Society, 1978.

Heiserman, David L. *Build Your Own Robot.* Blue Ridge Summit, Pa.: Tab, 1976.

Lucas, George. *Star Wars.* N.Y.: Ballantine, 1976.

Malone, Robert. *The Robot Book.* N.Y.: Harcourt, 1978.

Manchel, Frank. *An Album of Great Science Fiction Films.* N.Y.: Franklin Watts, 1976.

Shelley, Mary. *Frankenstein.* N.Y.: Signet, 1965.

Young, John F. *Cybernetic Engineering.* N.Y.: Wiley, 1973.

———. *Robotics.* N.Y.: Wiley, 1973.

You might also want to subscribe to *Robotics Age,* an advanced magazine on robots and robotics (P.O. Box 801, La Canada, Calif. 91011).

INDEX